Lecture Notes in Computer Science 4540

Commenced Publication in 1973
Founding and Former Series Editors:
Gerhard Goos, Juris Hartmanis, and Jan van Leeuwen

T0223313

Silvia Nittel   Alexandros Labrinidis
Anthony Stefanidis (Eds.)

GeoSensor Networks

Second International Conference, GSN 2006
Boston, MA, USA, October 1-3, 2006
Revised Selected and Invited Papers

 Springer

Volume Editors

Silvia Nittel
Anthony Stefanidis
University of Maine
Department of Spatial Information Science and Engineering
National Center for Geographic Information and Analysis
Orono, ME 044673, USA
E-mail: {nittel,tony}@spatial.maine.edu

Alexandros Labrinidis
University of Pittsburgh
Department of Computer Science
Advanced Data Management Technologies Laboratory
Pittsburgh, PA, 15260, USA
E-mail: labrinid@cs.pitt.edu

Library of Congress Control Number: Applied for

CR Subject Classification (1998): C.3, C.2.4, J.2

LNCS Sublibrary: SL 3 – Information Systems and Application, incl. Internet/Web
and HCI

ISSN   0302-9743
ISBN-10   3-540-79995-8 Springer Berlin Heidelberg New York
ISBN-13   978-3-540-79995-5 Springer Berlin Heidelberg New York

Springer is a part of Springer Science+Business Media

springer.com

© Springer-Verlag Berlin Heidelberg 2008
Printed in Germany

Typesetting: Camera-ready by author, data conversion by Scientific Publishing Services, Chennai, India
Printed on acid-free paper   SPIN: 12327874   06/3180   5 4 3 2 1 0

Preface

This volume serves as the post-conference proceedings for the Second GeoSensor Networks Conference that was held in Boston, Massachusetts in October 2006. The conference addressed issues related to the collection, management, processing, analysis, and delivery of real-time geospatial data using distributed geosensor networks. This represents an evolution of the traditional static and centralized geocomputational paradigm, to support the collection of both temporally and spatially high-resolution, up-to-date data over a broad geographic area, and to use sensor networks as actuators in geographic space. Sensors in these environments can be static or mobile, and can be used to passively collect information about the environment or, eventually, to actively influence it.

The research challenges behind this novel paradigm extend the frontiers of traditional GIS research further into computer science, addressing issues like data stream processing, mobile computing, location-based services, temporal-spatial queries over geosensor networks, adaptable middleware, sensor data integration and mining, automated updating of geospatial databases, VR modeling, and computer vision. In order to address these topics, the GSN 2006 conference brought together leading experts in these fields, and provided a three-day forum to present papers and exchange ideas.

The papers included in this volume are select publications, corresponding to extended versions of papers presented at the conference, and a few additional invited contributions; all papers went through a rigorous refereeing process. More information about the scientific background of geosensor networks in general and of the papers included in this volume in particular may be found in the Introduction chapter.

We greatly appreciate the many people who made this happen. Specifically, we would like to acknowledge the support of NSF, through the Sensor Science Engineering and Informatics (SSEI) IGERT program at the University of Maine (DGE-0504494), and especially Kate Beard, the principal investigator. We would also like to thank the University of Maine and the University of Pittsburgh for their support. From the University of Pittsburgh, we would especially like to thank George Klinzing, the Vice Provost for Research for his support of the GSN 2006 conference. We would also like to acknowledge the support of StreamBase. Last, but not least, we would like to thank everybody that helped in the organization of the GSN 2006 conference and the production of this volume. In particular, we would like to thank Blane Shaw, the authors and the participants of the conference, the Program Committee members and the Springer staff for their help.

March 2008

Silvia Nittel
Alexandros Labrinidis
Anthony Stefanidis

Workshop Organization

Organizers

Silvia Nittel, University of Maine, USA
Alexandros Labrinidis, University of Pittsburgh, USA
Anthony Stefanidis, George Mason University, USA

Program Committee

Karl Aberer, EPFL, Switzerland
Kate Beard, University of Maine, USA
Alastair Beresford, University of Cambridge, UK
Phillippe Bonnet, DIKU, Denmark
Ugur Cetintemel, Brown University, USA
Isabel Cruz, University of Illinois at Chicago, USA
Sylvie Daniel, Universite Lavel, Canada
Amol Deshpande, University of Maryland, USA
Matt Duckham, University of Melbourne, Australia
Dina Goldin, University of Connecticut, USA
Mike Goodchild, UC Santa Barbara, USA
Joe Hellerstein, UC Berkeley, USA
Christopher Jaynes, University of Kentucky, USA
Vana Kalogeraki, University of Riverside, USA
Nick Koudas, University of Toronto, Canada
Antonio Krueger, University of Muenster, Germany
Lars Kulik, University of Melbourne, Australia
Sam Madden, MIT, USA
Allan McEachren, Penn State, USA
George Percivall, Open Geospatial Consortium, USA
Dieter Pfoser, CTI, Greece
Mirek Riedewald, Cornell University, USA
Simonas Saltenis, Aalborg University, Denmark
Jochen Schiller, Free University Berlin, Germany
Mubarak Shah, University of Central Florida, USA
Cyrus Shahabi, USC, USA
Yoh Shiraishi, University of Tokyo, Japan
Andrew Terhorst, CSIR Satellite Applications Centre, South Africa
Yoshito Tobe, Tokyo Denki University, Japan
Niki Trigoni, Birkbeck, University of London, UK

Agnes Voisard, Fraunhofer ISST and FU Berlin, Germany
Peter Widmayer, ETHZ, Switzerland
Stephan Winter, University of Melbourne, Australia
Jun Yang, Duke University, USA
Vladimir Zadorozhny, University of Pittsburgh, USA
Guoqing Zhou, Old Dominion University, USA

Table of Contents

Applications

Introduction to Advances in Geosensor Networks

Silvia Nittel[1], Alexandros Labrinidis[2], and Anthony Stefanidis[3]

[1] Department of Spatial Information Science and Engineering
University of Maine Orono, ME 04469
nittel@spatial.maine.edu
[2] Department of Computer Science
University of Pittsburgh
Pittsburgh, PA 15260
labrinid@cs.pitt.edu
[3] Department of Geography and Geoinformation Sciences
George Mason University
Fairfax, VA 22030
astefani@gmu.edu

Advances in microsensor technology as well as the development of miniaturized computing platforms enable us to scatter numerous untethered sensing devices in hard to reach terrains, and continuously collect geospatial information in never before seen spatial and temporal scales. These geosensor network technologies are revolutionizing the way that geospatial information is collected, analyzed and integrated, with the geospatial content of the information being of fundamental importance. Analysis and event detection in a geosensor network may be performed in real-time by sensor nodes, or off-line in several distributed, in-situ or centralized base stations.

A large variety of novel applications for geosensor networks have already emerged. For example, real-time event detection of toxic gas plumes in open public spaces is crucial for public safety, while monitoring the progression of an oil spill is environmentally valuable. At the same time, more sophisticated applications are becoming feasible for the first time, for example, people using mobile phones equipped with microsensors and short-range communication to collect, exchange, and analyze local information with each other, whether it be about the freshness of produce in a supermarket, or the presence of influenza viruses in the air, or the presence of explosives at an airport. Geosensor networks may find applications in diverse fields, such as environmental monitoring (e.g., habitat observation and preservation, ocean and coastal monitoring), precision agriculture and fisheries, ad-hoc mobile computing for transportation, or surveillance and battlefield situations.

A geosensor network can loosely be defined as a sensor network that monitors phenomena in geographic space. Geographic space can range in scale from the confined environment of a room to the highly complex dynamics of an ecosystem region. Nodes in the network are static or mobile, or attached to mobile objects (e.g., on buses) or used by humans (e.g., cell phones). For example, cameras and GPS sensors on-board static or mobile small-form, mobile or stationary platforms have the ability to provide continuous streams of geospatially-rich information. Today, these types of geosensor networks can range in scale from a few cameras monitoring traffic to thousands of nodes monitoring an entire ecosystem.

S. Nittel, A. Labrinidis, and A. Stefanidis (Eds.): GSN 2006, LNCS 4540, pp. 1–6, 2008.
© Springer-Verlag Berlin Heidelberg 2008

Over the last 10 years, research efforts have taken place to develop the basic hardware infrastructure for small-scale sensor network systems consisting of large numbers of small, battery-driven sensor nodes that collaborate unattended and are self-organizing on tasks, and communicate via short-range radio frequency with neighboring nodes. Operating system software for these devices is available in the open source domain today, and the basic strategies of deploying sensor networks have a solid foundation.

Many challenges for geosensor networks still exist. The main challenge of sensor networks is to program the sensor nodes as a single, task-oriented computational infrastructure, which is able to produce globally meaningful information from raw local data obtained by individual sensor nodes. Another challenge is to integrate the sensor network platforms with existing, large-scale sensors such are remote sensing instrument or large, stationary ocean buoys, and process the information in real-time using a data streaming paradigm. Overall, the current research problems are centered at data management in general. Data management challenges for geosensor network applications can be divided into the areas of basic data querying, managing, and collection, which is today researched in the database community, and the more advanced, higher-level data modeling and formal data representation issues as well as data integration strategies, which are novel topics in spatial information science.

The spatial aspects of the overall technology are important on several (abstraction) levels of a geosensor network, as the concepts of space, location, topology, and spatio-temporal events are modeled on various abstraction levels. For example, the hardware and communication layers handle the physical space of sensor deployment, and communication topologies. The database layer generates execution plans for spatio-temporal queries that relate to sensor node locations and groups of sensors. Data representation and modeling deals with the relation between collected raw sensor data as fields and phenomena in geographic space.

The papers collected in this volume represent key research areas that are fundamental in order to realize the full potential of the emerging geosensor network paradigm. They cover the spectrum from low-level energy consumption issues at the individual sensor level to the high-level abstraction of events and ontologies or models to recognize and monitor phenomena using geosensor networks. We tried to cluster them across two separate research areas, namely *data acquisition and processing*, and *data analysis* and *integration*. This separation is to a certain extent arbitrary, as most papers address challenges that permeate across different research areas. Additionally, to better illustrate the complexity of transferring research into practice, we have also included three papers representing the diversity of geosensor network applications.

Section 1: Data Acquisition and Processing

The first section of this book is dedicated to papers that deal with *data acquisition and processing* challenges in geosensor networks, mostly on the sensor nodes themselves, or even within the sensor network. Since sensor networks depend on in-network processing to preserve energy and deal with the restricted communication bandwidth between sensor nodes, the challenge exists to come up with intelligent data collection strategies. Another challenge is the actual deployment strategy for sensor nodes, since they are likely scattered in the area of interest, but must be able to achieve network connectivity for all nodes. This also applies for mobile sensor networks, i.e. networks in which one, or more nodes are mobile.

Regarding *sensor deployment*, the paper of Ferentinos, Trigoni, and Nittel is addressing mobile sensor network data acquisition in a turbulent environment, where sensor nodes move involuntarily. The motivating application is ocean current tracking using a network of sensors that are drifting on the sea surface.

Energy management remains a crucial issue in sensor network deployment, with in-network communication being the major energy-consuming activity. Addressing this issue, the paper of Kulik, Tanin, and Umer is presenting a novel algorithm for queries in sub-network structures. Such queries need to reference only parts of a sensor network, thus offering opportunities to optimize data collection paths while minimizing energy consumption. Motivated by the same goal, namely the implementation of energy-efficient query processing in sensor networks, Deligiannakis and Kotidis present a comparative analysis of several data reduction techniques, ranging in scope from the simple monitoring of a node's variance to the identification of spatiotemporal correlations among nodes.

Processing the tsunami of data generated by sensor networks presents a host of well-known challenges that overwhelm classic database management systems, with their store-then-query processing paradigm. Data stream processing systems (DSPS) have been introduced to *handle* substantial volumes of data in a variety of applications, ranging from financial data to environmental monitoring. Typical queries in a DSPS are registered ahead of time; these queries are continuous, constantly being evaluated against a never-ending stream of incoming data to generate output streams. Accordingly, they are highly suitable for the processing requirements of sensor network applications. The paper of Tatbul, Ahmad, Cetintemel, Hwang, Xing, and Zdonik addresses the scalability and high availability aspects of a distributed stream processing system through the Borealis prototype. In particular, the authors discuss how to dynamically modify data and query properties of Borealis without disrupting the system's runtime operation, and how to adapt Borealis to tolerate changes and failures.

With the *current* trend of moving from spatial to spatio-temporal analysis, *mobility* is increasingly being seen as a first-class citizen in sensor networks. Addressing this issue, the paper by Agouris, Gunopulos, Kalogeraki, and Stefanidis introduces a spatio-temporal framework to support object and sensor mobility, inspired by active surveillance applications using optical sensors (video and still cameras). The key challenge they address is how to track moving objects using a small network of sensors that are also mobile (e.g., cameras on-board unmanned aerial vehicles). In the process, the authors touch upon issues related to the modeling of spatiotemporal information (e.g., the movement of a car), the development of similarity metrics to compare spatiotemporal activities (e.g., the movement of two different cars), and the management of a network to optimally track moving objects (e.g., repositioning sensors in order to follow certain activities). Continuing in the same context, the paper by Bakalov and Tsotras presents a novel indexing scheme for streaming spatio-temporal data, and efficient algorithms for evaluating spatio-temporal trajectory join queries, which used to identify a set of objects with similar behavior over a query-specified time interval. This supports queries about previous states of the spatio-temporal stream, provides approximations for object trajectories, and supports incremental query evaluation.

Section 2: Data Analysis and Integration

The second section of this book is dedicated to papers that deal with *data analysis and integration* challenges in geosensor networks, focusing especially on issues like 3D visual analysis, geosensor webs and standardization/interoperability, and higher-level semantic modeling.

Picking up the thread that ended the first section of this book, with camera-based surveillance, we have the paper of Akdere, Cetintemel, Crispell, Jannotti, Mao, and Taubin on visual sensor networks for 3D sensing. It presents the concept of a "virtual view" as a novel query mechanism to mediate access to distributed image data in a video sensor network. The ultimate objective is to establish efficient and effective networks of smart cameras that can process video data in real time, extracting features and 3D geometry in a collaborative manner.

Today, many traditional, larger-scale sensor field stations are already in place. With ubiquitous wireless communication networks, data can be retrieved via satellite links in real-time. Thus, the overall data collection paradigm in sensor data management is changing to a real-time scenario, and ultimately sensor networks will be integrated with larger-scale field stations and remote sensing imagery. Calling this scenario the *Geosensor Web*, we envision that access to sensor data will be as uniform and easy as access to data on the World Wide Web today. However, several problems exist that have to be addressed first before enabling such a vision. The article by Agrawal, Ferhatosmanoglu, Niu, Bedford, and Li focuses on the first challenge for sensor data integration, i.e. a framework comprising of real-time data streams from live sensors, a *stream-based middleware* for on-the-fly sensor data integration and analysis, linking it with ontologies and stored domain knowledge. Their driving application area is coastal forecasting and change analysis.

Viewing practical interoperable sensor data integration, industry-supported standards for access and sensor data presentation protocols are key. The paper by Botts, Percivall, Reed, and Davidson presents the Open Geospatial Consortium's (OGC) standard with regard to the standardized architecture for Sensor Web Enablement. Today, this standard is used for interoperable access to remote sensing instruments; however, it will become a highly important, enabling mechanism for the overall Geosensor Web. Many open research problems can still be found in the area of real-time sensor data integration such as novel meta data, access rights, copyright/privacy issues, uniform scale representation, scalability, and others.

Looking at the flood of collected and integrated real-time sensor data, it becomes clear that the cognitive aspects of users must be addressed and that higher-level, semantically rich data representation models and query languages are necessary. Users need to be able to express higher-level events such as "Track the toxic cloud, and report any topological changes" easily. This type of data representation model is also necessary to integrate sensor data with available domain knowledge and/or historic data.

Another mechanism for high-level and interoperable data representation of low-level sensor data are ontologies. Hornsby and King investigate the supporting role of ontologies for geosensor network data. In particular, they explore methods to link ontologies with real-time geosensor networks in order to augment the collected data with generalization or specialization relations from an ontology. For example, a geosensor network can observe moving cars on a freeway, tracking their location via a car identifier. The car identifier links the car to a classification scheme stored in a database; here, vehicles can be classified as military support trucks or medical support vehicles. Hornsby and King have implemented a mechanism to associate data values from the geosensor database with classes in the ontology, and support generalization or specialization queries based on the ontology.

Section 3: Applications

Today, the applications for geosensor networks are appearing in different domains ranging from habitat monitoring, watershed management, environmental pollution monitoring, deep sea explorations to monitoring food safety in South Africa and precision agriculture for large vineyards in Southern Australia. One can observe that a mindset change in the application areas is taking place. Scientists as well as practitioners are aware of technology advancements, which provide means for real-time availability of observational data and allowing existing sensor platforms to be networked. Networking sensor platforms of different types and scale provide an increased capability to correlate spatio-temporal information covering an entire region of interest. This paradigm shift from post-event, estimation based, historic data analysis to real-time, sensor-rich event detection and monitoring is fundamental in environmental applications. Adding small-scale geosensor network technology will change the awareness of the potential data scale over the next decade.

The paper by Pettigrew, Roessler, Neville, and Deese presents the GoMOOS project in the Gulf of Maine. Currently, the long-running coastal observing project consists of several, large-scale buoys distributed in the Gulf of Maine. Each buoy has several surface and underwater sensors attached, and uploads collected information via satellite link to a central computer. This information is used to model sea surface currents using a neural network approach since the available information currently is point based, thus, the currents in large areas need to be coarsely estimated. This project also relates to the paper by Trigoni, Ferentinos and Nittel in the second section discussing the deployment of a network of untethered ocean drifters to investigate the ocean currents directly, and potential collection information about nutrients and algae.

The next paper, by Terhorst, Moodley, Simonis, Frost, McFerren, Roos and van den Bergh, describes the deployment of another *environmental monitoring application*, targeted at detecting vegetation fires over Africa. Vegetation fires that burn over high-voltage electricity transmission lines can create line faults, which can disrupt regional electricity supply. Improving detection response time (ideally, making it real-time) can help mitigate the impact of such wild fires. Terhorst et al present the architecture of their Advanced Fire Information System and illustrate how a combination of two separate satellite systems with different characteristics can lead to improvement in detection accuracy.

Another interesting application area of geosensor networks is intelligent transportation systems. Location-based techniques have been used for the last decade to acquire information about the surroundings of the current location of a car or a pedestrian. Assuming the existence of sensors on devices as well as the ability for short-range, ad-hoc communication of nodes in close spatial proximity, a new range of application becomes possible. For example, several nodes collecting sensor information about their environment can collaborate and exchange, aggregate and forward this information. Often, sensed information is mostly relevant in the immediate neighborhood, and becomes less relevant with increasing distance. Wu, Winter and Guan address an approach to ride sharing using ad-hoc geosensor networks. Transportation clients, e.g. pedestrians, needing a ride from location l_1 to location l_2, request offers from transportation providers nearby. Transportation or ride providers can be private automobiles, taxis or public transportation. Based on the offers for complete or partial trips and accounting for the presence of competing clients, a client node has to strategize on accepting offers to cover the route at hand.

Outlook and Open Issues

The papers collected in this volume clearly demonstrate the interdisciplinary nature of geosensor networks, and their rapidly emerging potential to revolutionize the way in which we observe the physical world. There is an increased realization of the need for accurate and continuous monitoring of our environment. This environment is characterized by its inherent complexity (e.g., with the evolution of an ecosystem and the corresponding climate dynamics affecting and affected by human activities) thus mandating the realization of the potential offered by geosensor network applications.

If we attempt to identify relevant topics that have not yet emerged to the prominence that they deserve in our community, we can start with the issue of *sensor data* privacy. With the requirements to design ultra-light wireless communication protocols for small-form devices, there is limited room left for advanced encryption schemes. A related issue is the need for *authentication* of sensed data. If sensor networks are deployed in security-sensitive areas, built-in mechanisms need to be available to provide for such data authentication. A third open issue is *data quality*. Mechanisms need to assure that defective or incorrectly calibrated sensors are excluded from the computation, and that calibration is established individually as well as collectively before deployment and also continuously later on. Today, many research efforts in sensor networks are conducted under assumptions derived from the constraints of current hardware platforms such as the Berkeley motes. Many of these assumptions such as using radio broadcasting as communication modality or restricted battery life might not be valid anymore in a few years, and these assumptions might change completely. Lastly, one observation we can make based on the current state-of-the-art is that existing approaches to geosensor networks are rather passive: sensors observe events, and just report/record information. As geosensor networks become more mature and more pervasive, we expect the level of interaction between humans, geosensor networks, and the environment to increase dramatically.

Given the many different aspects and challenges of this interdisciplinary field, a single volume cannot possibly exhaust the emerging research agenda, but we believe that the papers collected here offer a valuable snapshot of current research, as it was reflected at the GSN'06 conference. We hope that this volume will serve as a reference point for new scientists venturing in this area, and hope that we will have them participate in future GSN conferences.

Data Acquisition and Processing

Impact of Drifter Deployment on the Quality of Ocean Sensing

Konstantinos P. Ferentinos[1], Niki Trigoni[2], and Silvia Nittel[3]

[1] Agricultural University of Athens, Athens, Greece
kpf3@cornell.edu
[2] University of Oxford, Oxford, UK
niki.trigoni@comlab.ox.ac.uk
[3] University of Maine, Orono, USA
nittel@spatial.maine.edu

Abstract. Traditional means of observing the ocean, like fixed moorings and radar systems, are expensive to deploy and provide coarse-grained data measurements of ocean currents and waves. In this paper, we explore the use of an inexpensive wireless mobile ocean sensor network as an alternative flexible infrastructure for fine-grained ocean monitoring. Surface drifters are designed specifically to move passively with the flow of water on the ocean surface and they are able to acquire sensor readings and GPS-generated positions at regular intervals. We view the fleet of drifters as a wireless ad-hoc sensor network with two types of nodes: i) a few powerful drifters with satellite connectivity, acting as mobile basestations, and ii) a large number of low-power drifters with short-range acoustic or radio connectivity. We study connectivity and uniformity properties of the ad-hoc mobile sensor network. We investigate the effect of deployment strategy. The objective of this paper is to address the following challenge: how can we trade the usage of resources (e.g. number of drifters, and number of basestations vs. communication range) and which deployment strategy should be chosen (e.g. grid-like, star-like, etc.) to minimize energy costs, whilst satisfying application requirements for network connectivity and sensing density. Using simulation and real dataset, we investigate the effects of deploying drifters with regard to the following questions: i) where/when should drifters be placed initially? ii) how many drifters should initially be deployed?, iii) the effect of the number of basestations (drifters with satellite connectivity) on the overall network connectivity, and iv) the optimal communication range of the basic drifters. Our empirical study provides useful insights on how to design distributed routing and in-network processing algorithms tailored for ocean-monitoring sensor networks.

1 Introduction

Marine microorganisms such as phytoplankton are exceedingly small (2-3 mm), and coastal ocean currents are a significant factor influencing the transport and circulation of these marine microorganisms. Establishing a detailed model of

S. Nittel, A. Labrinidis, and A. Stefanidis (Eds.): GSN 2006, LNCS 4540, pp. 9–24, 2008.
© Springer-Verlag Berlin Heidelberg 2008

ocean currents in a relatively small area such as a bay or a specific coastline is important since currents carry nutrients and potential toxins, which affect ecosystems and humans in coastal regions. For example, currents can distribute the Alexandrium fundyense algae during the warm summer months. This type of algae contaminates shellfish, and can make people who eat the shellfish seriously ill [7]. Today, the knowledge of the behavior of ocean and coastal currents is limited, and predicted trajectories of armful algae imbedded in the flow are inadequate from the point of view of public health.

Today, major ocean currents are established using coastal radar; however, the information is spatially and temporally coarse, and the gaps need to be filled in by interpolation techniques. In this paper, we investigate the alternative deployment of a fleet of inexpensive, networked ocean drifters, which are passively propelled by the currents and report their GPS-based location and trajectories to the end user.

Today's available drifter platforms such as [4] are deployed in a singular fashion, and use expensive satellite communication to upload data to a centralized computer. Satellite communication is high cost, from the standpoint of the equipment price, with regard to energy consumption and also the service fee for the satellite link service. In this paper, we explore the use of a fleet of inexpensive drifters as an alternative flexible infrastructure for fine-grained ocean monitoring. We view the fleet of drifters as wireless ad-hoc sensor network with two types of nodes: i) only a few powerful drifters with satellite connectivity, acting as mobile basestations, and ii) a large number of low-power, inexpensive drifters with short-range acoustic or radio connectivity. Our objective is twofold: using a fleet of small-scale sensor nodes that communicate with each other using lower-energy acoustic signals instead of a satellite uplink saves large amounts of energy. Additionally, the fleet provides more detailed information by covering an ocean region in high density. The passive movement of drifters can be used to derive actual ocean currents on a detailed scale.

Deploying such a fleet of mobile ad-hoc sensor nodes on the ocean surface is a novel research problem, both from the perspective of computer science and oceanography. We have explored communication connectivity and sensing uniformity of a fleet of a mobile ad-hoc sensor network using real datasets from the Liverpool Bay. The challenge is to design, build and deploy drifter platforms that despite involuntary, passive movement over long time periods (up to 3 months) preserve energy power, long-term network connectivity, and sensing uniformity [5]. The objective of this paper is to address the following challenge: how can we trade the usage of resources (e.g. number of drifters, and number of basestations vs. communication range) and which deployment strategy should be chosen (e.g. grid-like, star-like, etc.) to minimize energy costs, whilst satisfying application requirements for network connectivity and sensing density. Using simulation and real dataset, we investigate the effects of deploying drifters with regard to the following questions: i) where/when should drifters be placed initially? ii) how many drifters should initially be deployed?, iii) the effect of the number of basestations (drifters with satellite connectivity) on the overall network connectivity, and iv) the optimal communication range of the basic drifters. Our empirical study provides useful insights on how to design distributed routing and in-network processing algorithms tailored for ocean-monitoring sensor networks.

The remainder of the paper is organized as following: in Section 2, we provide relevant technical background on the current state of the art of ocean sensor networks, drifter platforms and wireless communication technology for water environments. In Section 3, we explore the research questions and the approach of this paper in more detail. Section 4 and Section 5 contain our experimental results and we conclude with Section 6.

2   Background

In this section, we provide background information about the current state of the art in ocean observation research. The research can be roughly divided into deep sea exploration using submarines and ocean bottom sensor platforms and robots connected by optical fiber cable (e.g. NEPTUNE [6]). Another main research domain is in near-coastal ocean observations using fixed, large moorings equipped with sensor and satellite connection for data upload. These sensor environments are extended with coastal radar stations, autonomous gliders, and research vessels. Our interest in ocean surface drifters is with regard to near-coastal deployments.

2.1   Ocean Drifters

Today, several projects and platforms for shallow water drifters exist. Initial larger-scale deployments of drifters were in the Gulf of Mexico and the Southwestern Caribbean Sea designed to explore the Gulf Stream in more detail.

Today, the international ARGO project [4] is one of the largest deployments of drifters in the world oceans. ARGO is an program that began in 2000, and now the deployment of 3000 profiling drifters is 100% complete. The purpose of ARGO is to examine the global ocean currents, circulation and air-sea interaction, with the goal of improving climate models and predictions.

The Argo drifter (also called "Davis Drifter") was designed to be a surface level (1 meter below surface) drifter which can report position via the Argos satellite link-based data collection system. Location determination by GPS is also available. The unit has a nominal operating life of 9 months. The global array of 3,000 floats is distributed roughly every 3 degrees (300km).

Currently, drifters are deployed in a singular fashion, and each drifter reports data via expensive satellite uplink instead to other drifters or data mules (such as gliders, buoys or ships). The topic of fleets of surface level drifters using inexpensive acoustic, radio or optical communication is today a interesting, novel research topic.

Networks of mobile wireless sensor nodes are currently investigated in deep sea applications such as the NEPTUNE project or in small-scale surface deployments. For example, the Starbug Aquaflecks and Amour AUV, developed by

MIT, are an underwater sensor network platform based on Fleck motes developed jointly by the Australian Commonwealth Scientific and Research Organization (CSIRO) and MIT CSAIL [11]. The 4 inch long Aquaflecks are combined with a mobile Amour AUV, which acts as a data mule to retrieve data from the different sensor nodes. The Amour AUV uses two types of communication, i.e. an acoustic modem for long-range communication and optical-modem for short range. The acoustic modem has a data rate of 220 bits/s over 5 km, while the optical modem has a throughput of 480bit/s with range of over 200m consuming 4.5mJ/bit.

2.2   Wireless Communication for Ocean Environments

Typically, underwater sensor nodes are connected to a network's surface station which connects to the Internet backbone through satellite communication or an RF link. The sensor nodes located in shallow or surface waters use diverse wireless communication technologies such as radio, acoustic, optical or electromagnetic signals. The different technologies vary with regard to communication *range* of the sender, *data rate* per second (data propagation speed), *energy consumption* and *robustness* with regard to noise or interference (such as Doppler effects) [1].

Radio signals are used in shallow water sensor networks, however, the travel speed of radio signals through conductive sea water is very low, i.e. about at a frequency of 30-300Hz. Experiments performed at the University of Southern California using Berkeley Mica2 Motes have reported to have a transmission range of 120 cm in underwater at a 433MHz radio transmitter [12]. Optical waves do not suffer from such high attenuation but are affected by scattering. Also, transmission of optical signals requires high precision in pointing the narrow laser beams, which is less practical in water.

Basic underwater acoustic networks (UWA) are the most commonly used communication media for water-based sensor networks [2,8]. Acoustic communication is formed by establishing two-way acoustic links between various sensor nodes. UWA channels, however, differ from radio channels in many respects. The available bandwidth of the UWA channel is limited, and depends on both range and frequency; the propagation speed in the UWA channel is five orders of magnitude lower than that of the radio channel. UWA networks can be distinguished into very long range, long, medium, short and very short communication range. As a rule, the shorter the communication range, the higher the bit rate. Typical ranges of acoustic modems vary between 10km to 90km in water. Furthermore, acoustic networks can be classified as horizontal or vertical, according to the direction of the sound wave. There are also differences in propagation characteristics depending on direction. Furthermore, acoustic signals are subject to multipath effects [10], large Doppler shifts and spreads, and other nonlinear effects.

Acoustic operation is affected by sound speed. Overall, the bit rate in water is about five orders of magnitude lower than in-air transmission. Sound speed is slower in fresh water than in sea water. In all types of water, sound velocity is

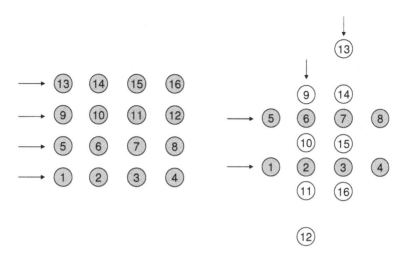

Fig. 1. Grid and hash deployments of drifters

affected by density (or the mass per unit of volume), which in turn is affected by temperature, dissolved molecules (usually salinity), and pressure. Today, the desired information transmission rate in the network is 100bit/s from each node. The available (acoustic) frequency band is 8-15 kHz. Uncertainty about propagation delays is typical of acoustic communication. Information is transmitted in packets of 256 bits, and nodes transmit at most 5 packets/h. Typical deployment of nodes can be as drifters or mounted on the ocean bottom, and separated by distances of up to 10km [9].

2.3   Data Management for Ocean Sensor Networks

Drifters are deployed to continuously collect data. At minimum, the end user is interested in the trajectory of the drifter itself since it contains relevant information about the ocean dynamics in the area covered. Furthermore, drifter platforms can be equipped with diverse sensors to sample the water. Today, salinity and temperature sensors are the most commonly used sensors. Drifter platforms can also carry accelerometers to measure wave speed or water acceleration for tsunami detection. Biological sensors detect marine microorganisms such as algae species and distribution.

Currently, drifters sense, store, and aggregate data locally until it is uploaded once a day via satellite connection to a centralized computer. Today, point sampling is common; region sampling via several collocated drifters during the same time period is rare. Typically, local data logger applications are run that contain limited processing and computing intelligence. Data collection is file-based, and reported in batch mode.

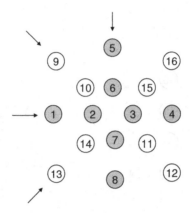

Fig. 2. Star deployment of drifters

3 Problem Description

Model assumptions: Consider a set of n drifters $D = \{d_1, \ldots, d_n\}$ deployed in the ocean to monitor a coastal area of interest. Let (t_i, x_i, y_i) be the time and location of initial deployment of drifter d_i. Drifters are designed to be passively propelled by local currents, are location-aware (using GPS), and are equipped with a variety of sensor devices to monitor different properties of the ocean surface. All drifters have local wireless communication capabilities that allow them to exchange messages with other drifters within range R [1]. A subset of the drifters ($B \subseteq D$) also has satellite connectivity, which allows them to propagate sensor data to oceanographers and other interested users around the globe. We refer to these special-purpose drifters as *mobile base-stations*, or simply *base-stations*. We thus view the set of drifters as a hierarchical mobile ad hoc network, wherein simple drifters forward their readings hop-by-hop to one of the mobile base-stations.

In order to predict drifter movement, we use a dataset CUR of coarse-grained radar measurements of current speed and current direction. Radar measurements are taken at regular intervals (e.g. every 1 hour) at various junction points of a grid spanning the area of interest (e.g. one pair of (speed,direction) measurements per $4km \times 4km$ grid cell). Current speed and direction conditions at all other locations are estimated using spline two-dimensional interpolation. Based on these current speed and direction measurements, we evaluate drifter locations over time, and we use the resulting trajectories as input to our simulations. In a real setting, drifter trajectories would be derived directly via GPS.

---

[1] In reality, the communication range is not a perfect circle, and the delivery ratio depends not only on the distance, but on a variety of environmental conditions. We leave the study of realistic communication models in ocean environments for future work.

Basestations	Grid Deployment	Hash Deployment	Star Deployment
1			
2	1,11	1,15	1,11
4	1,6,11,16	1,7,12,14	1,6,11,16
8	1,3,6,8,9,11,14,16	1,3,6,8,10,12,13,15	1,3,6,8,9,11,14,16

Fig. 3. Identifiers of drifters selected as basestations in the grid, hash and star deployments

Metrics: In this paper, we focus on empirically quantifying two aspects of drifter behavior: multi-hop connectivity and sensing coverage.

Multi-hop connectivity: This is defined as the percentage of drifters that can reach at least one of the base-stations on a multi-hop path. Multi-hop connectivity is useful for quantifying the ability of drifters to relay their readings hop-by-hop to the end-users through one or more base-stations.

Sensing coverage: We use two metrics of sensing coverage: i) *sensing density*, which is the number of connected drifters with multi-hop connectivity within the area of interest and ii) *sensing uniformity*, which denotes whether drifters are uniformly dispersed in the area of interest or congested in a small part of it. To quantify sensing uniformity, we adopt the definition of MRD (Mean Relative Deviation) proposed by Ferentinos and Tsiligiridis [3]:

$$MRD = \frac{\sum_{i=1}^{N} |\rho_{S_i} - \rho_S|}{N\rho_S}$$

where N is the number of equally-sized overlapping sub-areas that the entire area of interest is divided into. Sub-areas are defined by four factors: two that define their size (length and width) and two that define their overlapping ratio (in the two dimensions). In the formula above, ρ_{S_i} is the spatial density of connected drifters within sub-area i and ρ_S is the spatial density of connected drifters in the entire area of interest. Thus, MRD is defined as the relative measure of the deviation of the spatial density of drifters in each sub-area from the spatial density of drifters in the entire area of interest. Perfect uniformity (MRD=0) is achieved when each sub-area has the same spatial density as that of the entire area of interest, while higher MRD values correspond to lower uniformity levels of drifters.

Main objectives: Given the set of drifters D deployed at specific times and locations, the subset of base-stations B and a real dataset of current information CUR that determines drifter trajectories, we would like to address the following two questions:

– How does the deployment strategy, which is defined by the positions and timestamps of drifters when they are first deployed, impact the communication connectivity and sensing coverage of the fleet of drifters? In particular, our goal is to determine whether grid-like drifter formations are preferred to star-like drifter formations. We study this problem in Section 4.

– In a specific application framework, engineers will be faced with the question of how to balance network resources to achieve a certain density of connected drifters in an area of interest. They will have the option of varying the number of drifters, the number of basestations or the communication range between drifters. In this paper, we present a study of how these three factors impact density of connected drifters (Section 5). Based on this study, engineers can make informed decisions about how to build a network of drifters, taking into account the prices of drifters, basestations, or communication devices with varying ranges.

Real Datasets: In order to empirically address the questions posed before, we considered a realistic scenario of deploying drifters in the Liverpool Bay (UK). We used real datasets of surface current measurements monitored in the coastal area, to infer how drifters would move under the influence of these currents.

The Liverpool Bay data has been provided by the Proudman Oceanographic Laboratory Coastal Observatory Project, and it was measured by a 12-16MHz WERA HF radar system, which has been deployed to observe sea surface currents and waves in Liverpool Bay. In our simulations, we use current direction and current speed data measured hourly at the center of cells of a 7×10 grid. The size of each grid cell is $4km \times 4km$, and thus the size of the monitored area is $28km \times 40km$. A smaller 5×5 grid is considered to be the *area of interest* (of size $20km \times 20km$); recall that among connected drifters we only count those in the area of interest to measure sensing density, as defined in Section 3. This is in contrast with multi-hop connectivity, which is the percentage of all connected drifters, in and out of the area of interest. Current speed and direction conditions at locations inside the grid (other than the grid junctions) are estimated using two-dimensional spline interpolation. Drifter locations are estimated every 5 minutes. Our simulations typically last for 1.5 days, which corresponds to 432 5-min time-steps.

4 Impact of Deployment Strategy

In our study, we compare the performance of three different deployment strategies. In all of them, we assume that four boats start deploying drifters simultaneously, at time t=0, and in 1-hour intervals. Four drifters are deployed per boat (at times t, t+1hr, t+2hr and t+3hr). The initial positions and routes of the four boats vary across different strategies:

– *Grid deployment:* Four horizontal lines (left deployment in Figure 1)
– *Hash deployment:* Two horizontal parallel lines and two vertical parallel lines (right deployment in Figure 1)
– *Star deployment:* Two pairs of perpendicular lines in a 45-degree rotation (four lines in star formation in Figure 2)

In all three deployments, boats scan the same square area of interest of size $20km \times 20km$. Each boat starts deploying drifters along a line as shown in

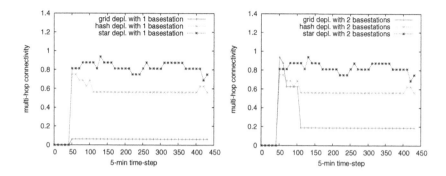

Fig. 4. Multi-hop connectivity with 1 and 2 basestations

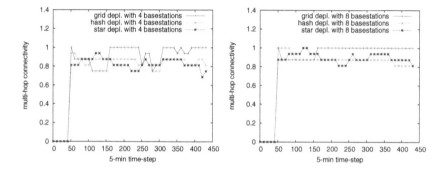

Fig. 5. Multi-hop connectivity with 4 and 8 basestations

Figures 1 and 2. The initial positions of the four boats, where the first four drifters are deployed, are illustrated by an arrow. It takes 1 hour for each boat to cross a distance of $4km$ where the next drifter is deployed.

Some of these drifters have satellite connectivity and act as mobile basestations. Table 3 shows which particular drifters are selected as basestations in scenarios with 1, 2, 4 and 8 basestations.

To compare the three deployment strategies we use the three metrics introduced in Section 3, namely multi-hop connectivity, sensing density and sensing uniformity (MRD: Mean Relative Deviation). We are interested in investigating which is the preferred deployment strategy in terms of connectivity and sensing, and how our decision may be affected by the number of basestations.

Multi-hop connectivity: Figures 4 and 5 show the behavior of the three deployment strategies as time elapses and as we increase the number of drifters acting as basestations. Notice that in scenarios with very few basestations, the grid deployment exhibits very low connectivity. In fact, only basestations manage to get readings through to the users using their satellite links, whereas all other drifters remain disconnected. The low connectivity is due to the fact that

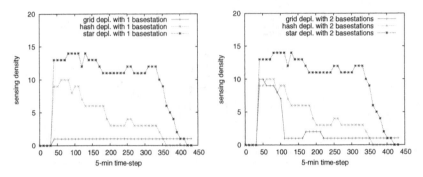

Fig. 6. Number of connected drifters in the area of interest (sensing density) with 1 and 2 basestations

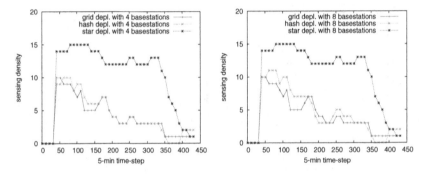

Fig. 7. Number of connected drifters in the area of interest (sensing density) with 4 and 8 basestations

the communication range (4 km) is equal to the distance between the locations of two consecutive drifters as they are initially deployed by a boat. Surprisingly the grid deployment exhibits very high multi-hop connectivity when 4 or 8 of the 16 drifters act as basestations.

The connectivity of the hash deployment is 10 times higher than that of the grid deployment in the case of 1 basestation, and 3 times higher in the case of 2 basestations. However, the relative difference between the two deployment strategies diminishes as we further increase the number of basestations (to 4 or more basestations).

The star deployment is shown to exhibit the highest multi-hop connectivity in most cases. Observe that the connectivity achieved with 1 basestation is very similar to that achieved with 8 basestations. This consistent behavior of the star deployment makes it very desirable, since it enables oceanographers to obtain readings from most drifters without using many expensive drifters with satellite connectivity (basestations). Before selecting the star deployment as the preferred strategy, we should first investigate whether it is equally efficient in terms of sensing density and sensing uniformity.

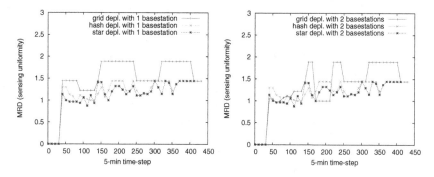

Fig. 8. Drifter uniformity in the area of interest (MRD) with 1 and 2 basestations

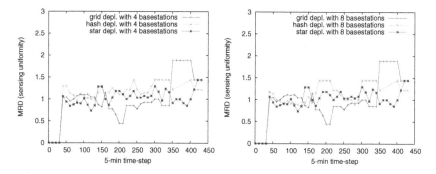

Fig. 9. Drifter uniformity in the area of interest (MRD) with 4 and 8 basestations

Sensing coverage: We examine two distinct metrics of sensing coverage, namely sensing density in Figures 6 and 7 and sensing uniformity in Figures 8 and 9. Recall that sensing density is the number of drifters within a region of interest that are connected through a multi-hop path to at least one of the basestations. In our experiments, the area of interest has size $20km \times 20km$, and each deployment has the same center as the area of interest.

Figures 6 and 7 show that for all deployment strategies, the number of connected drifters that remain within the area of interest decreases over time. It takes almost 400 time-steps ($400 \times 5 = 2000$ mins) for most drifters to be passively propelled by currents out of the area of interest.

In scenarios with few (1 or 2) basestations, the hash deployment yields higher density than the grid deployment. The gap, however, decreases significantly as we increase the number of basestations. The star deployment significantly outperforms the other two strategies in all cases, irrespective of the number of basestations. The density of drifters in the star deployment decreases more slowly than in the other two strategies up to time-step 300. Shortly after, we observe a very sharp descent of the density in the star deployment, until all drifters move outside the area of interest.

Observe that in scenarios with many basestations ($>= 4$), our results on multi-hop connectivity are very different from our results on sensing density. More specifically, in terms of multi-hop connectivity, the three deployment strategies have similar performance, whereas in terms of sensing density, the star deployment outperforms the others for prolonged time periods. This means that the star deployment places drifters so that most of them remain within the area of interest longer than if we had used grid or hash deployments.

The next question that arises is whether connected drifters remain uniformly dispersed in the area of interest, or whether they are clustered in small parts of the area. Figures 8 and 9 illustrate the performance of the three deployment strategies in terms of MRD (Mean Relative Deviation), which is a metric of sensing uniformity. The lower the MRD value the more uniform the spatial distribution of connected drifters. Recall that in scenarios with 1 or 2 basestations, the grid and hash deployments perform very poorly in terms of sensing density (Figures 6 and 7). Very few connected drifters are in the area of interest, and it is of little interest to compare the algorithms in terms of sensing uniformity. In scenarios with 4 and 8 basestations, the MRD values of the three deployment strategies fluctuate between values 0.5 and 2, with the grid deployment exhibiting slightly higher uniformity (lower MRD) than the other two approaches.

Summary of results: The star deployment exhibits significantly higher multi-hop connectivity than the hash and grid deployments in scenarios with few basestations (1 or 2 basestations out of 16 drifters). The three strategies behave similarly in scenarios with 4 or more basestations. The star deployment yields higher sensing density than the other two strategies in all cases. The grid deployment is slightly better in terms of uniformity, however, this benefit is too small to counteract the superiority of the star deployment in terms of sensing density. Our results are based on a dataset of real radar readings of current direction and velocity in the Liverpool Bay. They show that 1) the strategy used to deploy drifters has a significant impact on the network connectivity and sensing coverage, and 2) based on that dataset, the star deployment is preferred to the grid and hash deployments.

5   Resource Management

In the previous section, we compared three deployment strategies and selected the star deployment as the preferred approach. In this section, we examine variants of the star deployment in detail. Our aim is to understand the tradeoffs involved in varying the use of three network resources - number of drifters, number of basestations and communication range of simple drifters- to achieve a desirable sensing density. In each experiment, we fix the value of one of the three resources and measure the average sensing density as we vary the other two. Sensing density is averaged over 200 time-steps of 5 mins each (from time-step 100 to time-step 300).

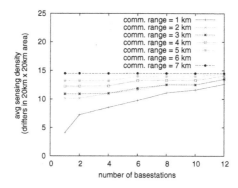

Fig. 10. Density of drifters in the area of interest as we vary the number of basestations and the communication range of simple drifters

We aim to address the following three questions:

1. How is sensing density affected by the number of basestations for various communication ranges? For example, if we want to increase the sensing density by δ, is it better to increase the communication range of drifters or convert some of the drifters into basestations by equipping them with satellite connectivity?
2. How is sensing density affected by the number of drifters for various communication ranges? If we want to increase network density by δ, is it better to increase the communication range of drifters of buy more drifters with the same communication range?
3. How is sensing density affected by the number of basestations for various numbers of drifters? If we want to increase network density by δ, is it better to increase the number of basestations or the number of drifters?

Figure 10 studies the tradeoff between the communication range of simple drifters and the number of basestations. The total number of drifters is set to 16. For small communication ranges, increasing the number of basestations is very beneficial in terms of sensing density. However, the benefit of adding a basestation is significantly decreased as we increase the communication range of simple drifters. Figure 10 shows that with 1 basestation and a 2 km communication range we observe an average of 10 drifters in the area of interest. To increase the sensing density from 10 to 14 drifters, we could either turn 10 drifters into basestations or we could increase the communication range of all drifters to 7 km. Depending on the price of each resource, network engineers can make an informed decision about how to reach their density goal.

Figure 11 studies the tradeoff between the total number of drifters and their communication range. The number of basestations is set to 2. Interestingly for small communication ranges (1-3km) the sensing density is initially sublinear in the number of drifters and then it becomes superlinear. Initially, when we insert a new drifter in the network, it might be disconnected from the other

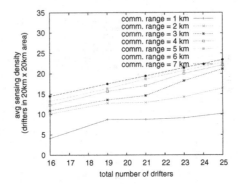

Fig. 11. Density of drifters in the area of interest as we vary the total number of drifters and their communication range

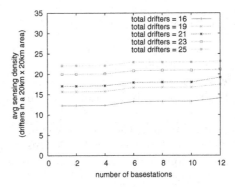

Fig. 12. Density of drifters in the area of interest as we vary the total number of drifters and the number of basestations

drifters because of the low communication range. However, as we keep inserting new drifters, density increases faster, because the new drifters connect not only themselves, but also previously disconnected drifters, to one of the basestations. Given a fixed number of basestations, if we want to increase network density, we have the option of increasing the number of drifters or their communication range. For example, Figure 11 shows that with 19 drifters and a 3 km communication range, we have on average 13 drifters in the area of interest. To increase the sensing density to 15 drifters we could either insert 3 more drifters or increase the communication range of all drifters to 4 km.

Figure 12 illustrates the tradeoff between the total number of basestations and the total number of drifters (sum of simple drifters and basestations). The communication range of simple drifters is set to 4 km. As we observed in Figures 6 and 7 of Section 4, the density of the star deployment is consistently high and it is not affected by the number of basestations. This is also confirmed in Figure 12, which shows that to increase network density, the only way is to increase the number of drifters.

6   Conclusions

In this paper, we investigated the impact of the drifter deployment strategy on the quality of ocean sensing performed using a fleet of drifters passively propelled by surface currents. These drifters form an adhoc sensor network; they monitor coastal waters with local sensor devices and propagate their readings hop-by-hop to a few mobile base-stations. Our comparison of the performance of three different deployment strategies, namely the grid, hash and star deployments, led us to the following conclusions.

The influence of deployment strategy over multi-hop connectivity depends on the number of basestations in the network. In cases with few basestations (1 or 2 basestations out of 16 drifters) the star deployment clearly exhibits significantly higher multi-hop connectivity than the hash and grid deployments. However, in cases with 4 or more basestations, all three strategies behave similarly. As far as sensing density is concerned, the star deployment outperforms the other two deployment strategies, irrespectively of the number of basestations in the network. In terms of uniformity of sensing points, the grid deployment is slightly better, however this benefit is too small to counteract the superiority of the star deployment in terms of sensing density. Thus, we conclude that 1) the strategy used to deploy drifters has a significant impact on the network connectivity and sensing coverage, and 2) the star deployment is preferred to the grid and hash deployments in the context of a network of drifters moving under real current conditions at Liverpool Bay.

Finally, we addressed three major tradeoff issues concerning ways to achieve specific increase in sensing density by an existing network of drifters. In the tradeoff between increasing the communication range of drifters or converting some of the simple drifters into basestations, we concluded that both ways are effective and the decision would be purely based on economical parameters. In the tradeoff between increasing the communication range of drifters or simply adding more drifters to the network, we concluded that for small communication ranges, increasing the communication range is more effective than adding more drifters, while for larger communication ranges both solutions are effective and again economical parameters come into play. Finally, in the tradeoff between adding more drifters to the network or converting some of the existing drifters into basestations, we concluded that the former approach is by far the most effective solution.

References

1. Akyildiz, I.F., Pompili, D., Melodia, T.: Challenges for efficient communication in underwater acoustic sensor networks. ACM SIGBED Review 1(2), 3–8 (2004)
2. Akyildiz, I.F., Pompili, D., Melodia, T.: Underwater acoustic sensor networks: research challenges. Ad Hoc Networks 3(1), 257–279 (2005)
3. Ferentinos, K.P., Tsiligiridis, T.A.: Adaptive design optimization of wireless sensor networks using genetic algorithms. Elsevier Computer Networks 51(4), 1031–1051 (2007)

4. Gould, J., et al.: Argo profiling floats bring new era of in situ ocean observations. EOS 85(19), 179–184 (2004)
5. Nittel, S., Trigoni, N., Ferentinos, K., Neville, F., Nural, A., Pettigrew, N.: A drift-tolerant model for data management in ocean sensor networks. In: MobiDE, pp. 49–58. ACM Press, New York (2007)
6. U. of Washington. The neptune project page
7. Pettigrew, N., Churchill, J.H., Janzen, C., Mangum, L., Signell, R., Thomas, A., Townsend, D., Wallinga, J., Xue, H.: The kinematic and hydrographic structure of the gulf of maine coastal current. Deep Sea Research II 52, 2369–2391 (2005)
8. Proakis, J., Sozer, E., Rice, J., Stojanovic, M.: Shallow water acoustic networks. IEEE Communications Magazine 39(11), 114–119 (2001)
9. Sozer, E., Stojanovic, M., Proakis, J.: Underwater acoustic networks. IEEE Journal of Oceanic Engineering 25(1), 72–83 (2000)
10. Stojanovic, M.: Recent advances in high-speed underwater acoustic communication. IEEE Journal of Oceanographic Engineering 21, 125–136 (1996)
11. Vasilescu, I., Kotay, K., Rus, D., Dunbabin, M., Corke, P.: Data collection, storage, and retrieval with an underwater sensor network. In: SenSys 2005: Proceedings of the 3rd international conference on Embedded networked sensor systems, pp. 154–165. ACM Press, New York (2005)
12. Zhang, B., Sukhatme, G., Requicha, A.: Adaptive sampling for marine microorganism monitoring. In: IEEE/RSJ International Conference on Intelligent Robots and Systems (2004)

Efficient Data Collection and Selective Queries in Sensor Networks

Lars Kulik, Egemen Tanin, and Muhammad Umer

National ICT Australia
Department of Computer Science and Software Engineering
University of Melbourne, Victoria, 3010, Australia
{lars,egemen,mumer}@csse.unimelb.edu.au

Abstract. Efficient data collection in wireless sensor networks (SNs) plays a key role in power conservation. It has spurred a number of research projects focusing on effective algorithms that reduce power consumption with effective in-network aggregation techniques. Up to now, most approaches are based on the assumption that data collection involves all nodes of a network. There is a large number of queries that in fact select only a subset of the nodes in a SN. Thus, we concentrate on selective queries, i.e., queries that request data from a subset of a SN. The task of optimal data collection in such queries is an instance of the NP-hard minimal Steiner tree problem. We argue that selective queries are an important class of queries that can benefit from algorithms that are tailored for partial node participation of a SN. We present an algorithm, called Pocket Driven Trajectories (PDT), that optimizes the data collection paths by approximating the global minimal Steiner tree using solely local spatial knowledge. We identify a number of spatial factors that play an important role for efficient data collection, such as the distribution of participating nodes over the network, the location and dispersion of the data clusters, the location of the sink issuing a query, as well as the location and size of communication holes. In a series of experiments, we compare performance of well-known algorithms for aggregate query processing against the PDT algorithm in partial node participation scenarios. To measure the efficiency of all algorithms, we also compute a near-optimal solution, the globally approximated minimal Steiner tree. We outline future research directions for selective queries with varying node participation levels, in particular scenarios in which node participation is the result of changing physical phenomena as well as reconfigurations of the SN itself.

1 Introduction

Efficient data collection and aggregation algorithms for sensor networks (SNs) exploit the fact that a sensor node consumes significantly less energy for information processing than for communication. Aggregating information at the node level such as computing the sum or the average of sensor readings reduces the need for communication: instead of transmitting the packets of each individual node separately, a node first aggregates the incoming packets of the nodes in communication range and then communicates the aggregated information to the next node in the collection path.

S. Nittel, A. Labrinidis, and A. Stefanidis (Eds.): GSN 2006, LNCS 4540, pp. 25–44, 2008.

Major in-network data processing techniques for SNs do not take an explicit position on the issue whether or not a query predicate selects all the nodes of a SN. Instead, most techniques implicitly assume that a query requires *all* nodes to respond. We refer to the sensor nodes that report their readings to a selective query as *participating nodes*. An example of a *selective query* is "SELECT the humidity readings FROM all sensors WHERE the temperature is above 40° for a DURATION of 2 hours EVERY 5 minutes". We call all nodes that fulfill the WHERE clause the participating node set of this query.

Current SN data management models such as Cougar and TinyDB view the SN as an ever growing relation(s) of tuples that are distributed across the sensor nodes [1,2]. They mimic classical relational database management systems. In classical systems query predicates limit the number of tuples that form the output relation. The query predicates in these models also serve to limit the set of sensor nodes that contribute to the answer of a query. Techniques developed for efficient data collection in SNs, in particular classical tree-based and multipath-based techniques [3,4], generate a nearly optimal number of messages for aggregation operations if all nodes need to report to a query. We show that these major in-network data processing schemes do not continue this optimal behavior for selective aggregate queries.

Most approaches for data collection, do not explicitly address query selectivity while computing an efficient data collection path. There are two main directions in SN query processing for optimizing the data collection process for selective queries: (1) preventing that a query is sent to nodes that do not fall into the scope of that query and, therefore, are not aware of the query and do not need to respond, and (2) minimizing the number of non-participating nodes in the collection path. An example for the first direction is the concept of semantic routing trees [5] where optimization between queries is the main focus. In this paper, we address the second direction. The main contribution of our research is a strategy that minimizes the number of nodes used in processing a query by discovering constrained regions to grow sub-trees for data collection and combining these trees in an efficient manner to transmit the final result to the sink. This strategy contrasts with earlier tree-based approaches, such as TAG, where the tree is created in a random manner using local greedy parent selection policies [1].

Our algorithm, called Pocket Driven Trajectories (PDT) is based on the insight that spatial correlation in sensor values coupled with query selectivity results in a set of participating nodes formed by one or more geographically clustered sets. We refer to these geographical clusters as *pockets*. The PDT algorithm first discovers the set of pockets for a given query and then aligns the aggregation tree to the spatially optimal path connecting these pockets. This path maximizes the use of participating nodes in the tree and conversely minimizes the number of non-participating nodes. The PDT algorithm is best suited to the selective queries that regularly collect data from a relatively consistent set of nodes over time. The initial set-up cost for the PDT algorithm can be amortized over the query lifetime.

The task to minimize the number of non-participating nodes in processing a query can be modeled as a minimal Steiner tree problem; which is known to be NP hard [6]. The number of nodes used in the globally approximated minimal Steiner tree for a given set of participating nodes can be seen as a near-optimal solution for the minimum number of nodes required for data collection. There are a number of algorithms to compute an approximation of the minimal Steiner tree efficiently [7,6,8]. These approximation algorithms, however, require global knowledge of the communication graph. This knowledge is typically not available in a SN. For a SN we require techniques that do not rely on global information about the network. The PDT algorithm can be considered as an approximation to the minimal Steiner tree that is solely based on local information. Our experiments show that for selective, aggregate queries, PDT does not only give better performance results than major in-network aggregation schemes but is also close to a globally approximated minimal Steiner tree.

A chief contribution of our work is the performance comparison of PDT and other major aggregation algorithms using extensive simulations in a number of node participation scenarios. We identify important parameters to capture the spatial properties of a node participation scenario and use these as a basis for our analysis. We show in our experiments that selectivity-awareness can reduce the usage of a large number of non-selected nodes in the data collection path, in particular, if the query selectivity is high.

Finally, we present future research directions where node participation is affected by changing physical phenomena as well as the SN configuration.

2   Related Work

Three major techniques for efficient data collection and aggregation are *packet merging*, *partial aggregation*, and *suppression*. The communication cost in wireless networks is determined by the number of packets a node has to transmit. Each packet has a fixed size and the size of a sensor reading, record, is typically much smaller. Packet merging [2] combines several smaller records into a few larger ones and thus reduces the number of packets and thereby the communication cost. Partial aggregation computes intermediate results at the node level such as the sum or the average of sensor readings. Suppression-based techniques only transmit values if the sensed values have changed from the previous transmission or differ from the values of neighboring sensor nodes [9]. Orthogonal techniques minimizing communication are compression techniques [10], topology control, or approximation of sensor readings.

TinyDB [1] enables in-network aggregate query processing using a generic aggregation service called TAG (Tiny AGgregation) [3]. TAG is one of the pioneering tree-based in-network aggregation schemes. To gather data within the SN, the sink is appointed to be the root of a tree and broadcasts its identifier and its level. All nodes that receive this message without an assigned level, determine their own levels as the level in the received message incremented by one. The identifier in the message determines the parent for the nodes receiving the message. In a lossless network in which all nodes are selected by a query, the resulting collection tree is close to an optimal solution. Aggregation in TAG is implemented by a merging function, an initializer, and an evaluator, and the aggregation operator is applied at every internal node.

In order to further optimize the data aggregation process, TinyDB introduces the concept of a semantic routing tree (SRT) [1,5]. Although the concept of selectivity is not explicitly addressed in TAG [3], SRTs can support selective queries. The basic idea behind SRTs is to proactively prune the SN in the query dissemination stage by ensuring that a selective query is sent only to those nodes that fall under its scope. SRT maintains meta-information at each internal node of the aggregation tree. More precisely, an SRT is an index over a fixed attribute A, for example the temperature sensed by the network, where each parent maintains the range of all values of its children for the attribute A. When a query is received by a parent, it forwards the query only when at least one child satisfies the predicate. An SRT optimizes the query forwarding phase of TAG and greatly reduces the number of broadcasts required to reach the nodes selected by the query. To avoid a high cost of maintenance, SRTs are designed for constant attributes such as the temperature or the location [1]. However, the maintenance cost of SRT can exceed its benefit if it has to be maintained for frequently varying attributes [1]. SRTs do not focus on the data collection optimization but on the broadcast of the query.

Tree-based aggregation schemes can be extended for changing network conditions [11]: aggregation operators are pushed down in an aggregation tree and adapt to changing conditions, such as a sub-tree that generates more readings than a sibling. This approach incrementally improves on existing schemes. In our work, we develop an aggregation scheme that, after retrieving the initial readings from a SN, specifically tailors the collection path to the sensor readings and the set of selected nodes in a SN.

Data collection paths are susceptible to link and node failures in a SN [4,12]. If a link or node fails that is close to the sink, the aggregated information of an entire sub-tree might be lost. Multi-path aggregation algorithms exploit the benefits of the wireless broadcast advantage that all nodes in communication range can hear a message and propagate the aggregates toward the sink using multiple routes. As a consequence data collection along multiple paths can be more robust for node failures or communication losses. In return, a multi-path aggregation algorithm has to cope with redundancy and deviations in data aggregation [13].

In [4] multi-path aggregation algorithms are seen as energy efficient as tree-based ones because each node only has to transmit a message once, in the same way as in any tree-based aggregation algorithm. However, a crucial assumption is that the receive time for each communicating sensor is not increased by multiple readings. Recent studies on the energy consumption of sensor nodes report that the energy requirement for the receive mode is only slightly lower compared to the energy cost for the transmission mode [14], i.e., a sensor node consumes almost the same amount of energy in receiving data as in transmitting data. In multi-path aggregation, each node expects to receive data from all nodes in communication range that have a higher level than the listening node. Therefore, the duration for receiving data can be considerably longer compared to a tree-based aggregation algorithm, where each node only has to listen to its direct children, a set that can be a significantly smaller. In addition, we show that with decreasing participation levels for a selective query, the energy cost for multi-path schemes can be quite high than a tree-based scheme or our PDT algorithm.

To overcome the higher energy costs, an approach that locally applies multi-path aggregation locally but not to the entire network is suggested in [15]. This approach is a hybrid aggregation scheme that combines the benefits of the two major aggregation schemes to form a more efficient option than pure multi-path aggregation schemes: a multi-path-based aggregation scheme is preferred to a simple tree-based aggregation scheme when the in-network aggregation operator is close to the sink; for deeper levels of aggregation tree, the operators work as if they are on a TAG-like aggregation tree as the loss of a sensor at deeper levels only marginally effects the final result.

A key challenge for multi-path routing schemes is to develop duplicate-insensitive algorithms for each aggregation operator. A naive approach could include meta information in each aggregated message such as the node identifier that participated in the creation of an aggregate, which could be used by forwarding nodes to suppress duplication. Although this approach could achieve the same accuracy as a tree-based aggregation algorithm, the limited storage and processing capabilities of sensor nodes, however, render such a scheme impractical for larger SNs. Thus, all multi-path schemes integrate much cheaper probabilistic Order and Duplicate Insensitive (ODI) methods of the sketch theory [4,13]. Therefore, current research in multi-path aggregation focuses on the development of better ODI algorithms to reduce the approximation error.

Data collection in SNs can be optimized using spatial and temporal suppression-based techniques [9]. Temporal suppression is the most basic method: the transmission of a sensor reading from a node is only allowed if its value has changed from last transmission. Spatial suppression includes methods such as clustered aggregation and model-based suppression [16]. They aim to reduce redundant transmissions by exploiting the spatial correlation of sensor readings. If the sensor readings of neighboring sensor nodes are the same, the communication of those sensed values can be suppressed. With CONCH [9], a hybrid spatio-temporal suppression algorithm, is introduced, that considers the node readings and their differences along the communication edges to suppress reports from individual nodes. Suppression is an orthogonal data flow optimization method to our approach and can easily integrated into other approaches.

Clustered in-network aggregation is a spatial suppression technique exploiting the spatial correlation of sensor readings to preserve energy [17,18,19]. Spatial correlation in sensed data refers to the fact that sensor readings in close proximity are typically similar. Spatial correlation is a frequent phenomenon for physical phenomena such as temperature or humidity [20]. If a selective query has to retrieve an aggregate such as the average temperature in a certain area, then nearby nodes typically have similar readings and are geographically clustered. Hence, only one node needs to respond to an aggregate query from a cluster [18,19] as in the Clustered AGgregation (CAG) and the Geographic Adaptive Fidelity (GAF) approaches. In static clustering [17], the network is statically partitioned into grid cells. For each grid cell one node is appointed as a cluster head that acts as a gateway but every node that has to respond to a query still reports its readings. In each cell, data is routed via a local tree, aggregated at the local gateway and then communicated to the sink. Methods that rely on approaches such as clustered in-network aggregation (such as CAG) have the disadvantage that the reported results can deviate from the real sensor readings. However, static clustering does not have this issue and is comparable to our work. Unlike our work, static clustering does not tune the collection paths to the specific network conditions.

In [21], the optimal data collection problem in SNs has been identified as a Steiner tree problem. The authors propose a data collection scheme based on a global Steiner tree approximation [8]. The main disadvantage of their approach is the requirement that each sensor node must have global knowledge in terms of network connectivity and minimum-hop routes, a knowledge that is costly and difficult to maintain in a SN where nodes have very limited capabilities. Each update of the global knowledge about such a graph at each node leads quickly to unnecessary storage and communication overheads. The PDT algorithm uses only local information that is easily available to any node in the network. Furthermore, the PDT collection path is query specific, i.e., no permanent information has to be kept or maintained.

3   A Location Based Aggregation Algorithm

Queries such as the example query in Section 1 that identify all locations with a temperature higher than 40° represent an important type of SN queries. In a fire monitoring system, this query could be used to collect data about the humidity levels in a large region in order to identify areas that have a severe fire risk. Queries of this type run for a period of time and periodically retrieve selected sensor readings from the SN. We refer to these queries as *selective recurrent queries*.

Selective queries, in contrast to an exhaustive or a snapshot query, can benefit from an optimized data collection strategy. In snapshot queries, where the sink cannot predict the number or the location of participating nodes in advance, it can be ineffective to spend resources on the discovery of the participating nodes. In exhaustive queries, where all nodes have to respond to a query, a random tree such as TAG often provides a near-optimal data collection strategy. However, in selective queries, we show that it is beneficial to invest resources for identifying nodes that do not need to be part of the data collection especially if the query involves many sampling periods, or epochs. The initial cost for an optimized aggregation path can later later be amortized during the life time of a query. Thus, selective recurrent queries are the primary motivation for the PDT algorithm.

In general, it is not possible to find a data collection and aggregation strategy that only employs those nodes that need to participate for a given query. Due to the constraints on the transmission range for sensor nodes, a data gathering algorithm usually has to include some nodes that are not selected by the query in order to reach the sink. The quality of a data aggregation scheme is determined by the total number of non-participating nodes used in a query. The smaller the number of non-participating nodes, the more energy efficient an algorithm will generally be. In the PDT algorithm, we minimize the number of non-selected nodes in the data collection structure by spatially restricting the aggregation path to a corridor that connects the pockets in an energy efficient manner.

The problem of reporting aggregates back to a sink by involving minimum number of non-participating nodes can be seen as an application of the minimum Steiner tree problem: given a graph $G = (V, E)$ and a set of terminal nodes $T \subset V$, we seek a minimum cost spanning tree that includes all terminal nodes [22]. In our case, the terminal nodes are the participating nodes. The minimum Steiner tree problem is known to be NP-hard and has been widely discussed in the literature [6].

In order to compare aggregation schemes with the optimal aggregation tree, the minimum Steiner tree, we outline a popular global Steiner tree approximation algorithm developed by Kou, Markowsky, and Berman [7] (henceforth referred to as KMB). The KMB algorithm allows us to compute a Steiner tree that has been shown to achieve a

mean efficiency that is not worse than 5% compared to the optimal Steiner tree [6,23].
For a graph G and a set of terminal nodes T, the KMB algorithm first computes a complete distance graph $G' = (V', E')$ for G such that $V' = T$ and the weight of each edge
e' in E' is the cost of the minimum cost path between the nodes of e' in G. Then, the
algorithm computes a minimum spanning tree MST' of G', translates MST' to G by
substituting every edge of MST' with the corresponding path in G, and finally removes
any possible cycles resulting from the translation.

The KMB algorithm is not directly applicable to SNs because, the algorithm requires
global knowledge of the node connectivity for any node in the graph. Therefore, we develop a *localized* aggregation scheme, called PDT (pocket driven trajectories), that approximates the minimal Steiner tree for a selective query. The experiments in Section 4
show that the resulting aggregation tree is comparable to KMB's global approximation.

3.1 Algorithm Overview

We assume that a SN with n nodes is represented by a connected unit disk graph $G = (V, E)$, where V is the set of sensor nodes and E the set of communication links. Each
query Q issued by a sink S selects a subset T of V. The PDT algorithm works as follows
(Figure 1):

1. the sink broadcasts the query Q and establishes a tree as in TAG;
2. during the first epoch the sink discovers a set of pockets $P = \{p_1 \ldots p_k\}$ that
   partitions the set T;
3. the sink computes a complete graph $G' = (V', E')$, where $V' = P \cup \{S\}$ and each
   edge weight is the Euclidean distance in the SN;
4. the sink computes the minimum spanning tree MST' of G';
5. the sink establishes an aggregation tree aligned to MST';

One of the key steps in the PDT algorithm is the localized discovery of pockets by the
sink. A pocket is a cluster of nodes selected by a query that are proximal, i.e., within a
certain distance. Due to spatial correlation, pockets are common while sensing physical
phenomena. We refer to the time between the two sampling operations of a query as an
epoch following the terminology used in data aggregation for SNs. In the following, we
describe a novel pocket discovery method, *location aggregation*, that computes pockets
with minimal overhead.

Figure 1 illustrates possible pockets that are selected by a query issued at a sink.
The sink broadcasts the query and all sensor nodes build a random query tree. In the
first epoch, participating leaf nodes start the aggregation phase by sending the requested
sensor readings to their parent nodes and by piggybacking their location information as
Euclidean coordinates. Parent nodes recursively apply the query and location aggregation operators and forward the partial aggregates to their parent nodes. The aggregation
operation proceeds as in classical data collection schemes. The key step is the location
aggregation, performed in parallel to data aggregation. The location aggregation operator merges locations to an enclosing rectangle if the children nodes are proximal leaf
nodes, and forwards the non-proximal nodes as singletons. If a participating node receives a rectangle, it merges the rectangle with its own position into a larger rectangle,

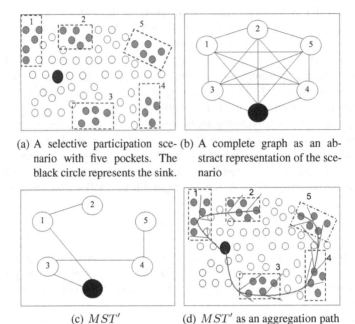

(a) A selective participation scenario with five pockets. The black circle represents the sink.

(b) A complete graph as an abstract representation of the scenario

(c) MST'

(d) MST' as an aggregation path

Fig. 1. The computation of the PDT for a selective participation scenario

if its position is proximal, and otherwise simply forwards the existing rectangle and singleton locations with its own location. Since the location information is piggybacked with the desired data, we expect that the location aggregation incurs a small overhead in terms of communication costs. Moreover, the successive merging of close pockets keep the volume of information small.

At the end of first epoch, the sink receives the queried aggregate and after applying location aggregation operation discovers the pockets $(p_1 \ldots p_k)$ selected by the query. The sink then computes a complete graph G' as explained in the overview of the PDT algorithm (see Figure 1(a), 1(b)). The algorithm then creates a minimum spanning tree MST' for G' at the sink. MST' is a pocket driven trajectory that optimizes aggregation for the specific pocket layout (Figure 1(c)). The PDT information can be encoded as a series of locations, each corresponding to either a sink location or the center point of one of the pocket rectangles. During the next epoch, the sink establishes the PDT by broadcasting the PDT information to all its direct children. All participating nodes that receive the PDT information packet join the trajectory by reassigning their parents to that node from which they receive the PDT information. Non-participating nodes decide to join the trajectory depending upon their distance to the trajectory and only nodes that join the trajectory forward the information packet. The successive forwarding and parent switching leads to a new aggregation tree aligned with the PDT (Figure 1(d)). This new tree is afterwards used for data forwarding and aggregation. The initial TAG tree is still maintained because in future epochs previously non-participating nodes may participate. Future participating nodes might have never heard the PDT information and thus have to use the original tree for data aggregation.

3.2 Shortcomings and Overheads

A query in a SN can consist of a large number of epochs. Even if the change per epoch is small, the node participation can change significantly during the lifetime of a query. The PDT algorithm is ideal if the change per epoch is relatively small so that the pockets do not change significantly in every step. Under those conditions, the PDT may have to be realigned a few times during the lifetime of a query. However, currently the PDT algorithm does not adapt to change within a query. We leave the investigation of more dynamic scenarios to future work.

The increase in packet size due to the aggregation of location information increases the communication overhead. The location information consists of aggregated pocket information and a list of atomic locations. Due to the spatial correlation of physical phenomena, the number of atomic locations is typically small and singletons mostly occur at the leaf level of the aggregation tree. The singletons are almost completely merged at the lower levels of the tree into pockets that traverse for the remainder of the tree in a compact form as rectangles. Our experiments show that the location information aggregation only slightly increases the communication messages.

The announcement of the PDT is the other overhead of the PDT algorithm. The number of extra messages generated during PDT information broadcast phase is equal to the number of nodes that decide to join the PDT. The initial setup cost can be amortized over the query lifetime. However, the initial cost cannot be amortized for snapshot queries and hence we do not recommend the use of the PDT algorithm for such queries.

4 Experimental Evaluation

4.1 Evaluation Parameters

In this section we compare the performance of the PDT algorithm and other major in-network aggregation schemes in a variety of SN settings. We first lay out the evaluation parameters that we use to analyze the impact of spatial characteristics of selective aggregate queries.

Spatial Selectivity Index. The nature and extent of spatial clustering of participating nodes may depend upon a number of factors, such as the magnitude of spatial correlation, SN configuration, the query predicate, and so forth. We introduce an integrated measure, called spatial selectivity index, SSI, that describes the extent of spatial clustering in a SN. SSI is based on two key parameters, *node scattering* and *pocket scattering*.

The *node scattering*, NS, at time t (we will drop the argument t for the sake of simplicity) describes the distribution of pockets for a query as the ratio of the total number of pockets P relative to the number of participating nodes N:

$$NS = P/N$$

The *pocket scattering*, PS, characterizes the degree of dispersion for the pocket locations. We define the *pocket centroid*, PC, of a pocket P_i as the sensor node that is closest to the average location of all nodes belonging to P_i. We then define the *global pocket centroid*, GPC, as the node that is closest to the centroid of all pocket centroids $P_1, \ldots, P_l$ partitioning the nodes selected by the query:

$$\text{GPC} = \frac{1}{l} \cdot \sum_{i=1}^{l} \text{PC}(P_i)$$

Let $\text{HC}(v, v')$ denote the minimum number of hops for a path connecting two nodes v and v'. Then, the pocket scattering PS is defined as the average of hop counts connecting the global pocket centroid with the pocket centroids of the pockets $P_1, \ldots, P_l$:

$$\text{PS} = \frac{1}{l} \cdot \sum_{i=1}^{l} \text{HC}(\text{PC}(P_i), \text{GPC})$$

We use the hop count between two nodes as a distance measure instead of their Euclidean distance. Since deployments of SNs can exhibit holes and barriers, the hop count provides a realistic approximation of the actual communication cost. It should be noted that the use of hop count as a distance measure is purely for evaluation purposes. The PDT algorithm itself does not use the hop count measure due to the practical limitations in making such information available locally at each node.

The spatial selectivity index SSI is then defined as a measure that describes the impact of node scattering as well as of pocket scattering for a given scenario:

$$\text{SSI} = \text{NS} \cdot \text{PS}$$

Lower SSI values characterize scenarios that are well pocketed and have a small pocket scatter, for example see Figure 3(a) and 3(b) in Section 4.3 that show two network deployments with different SSI values.

Sink Centroid Distance. We formalize sink position in order to analyze the affect of sink location on the performance of an aggregation algorithm. The ideal position of a sink is the GPC. The *sink centroid distance*, SCD, measures the hop count between the sink and its ideal position. If S is the position of the sink, the sink centroid distance SCD is defined as

$$\text{SCD} = \text{HC}(S, \text{GPC})$$

Co-Connectivity of a Deployment. Although a rectangular deployment is common in simulations, e.g., without any holes that alter connectivity, in this paper, we also consider the impact of irregular deployments on aggregation algorithms. Particularly, we measure the impact of holes. To simplify our discussion, we only take the total number of holes and their normalized size into account. If $|\mathcal{R}|$ denotes the size of the deployment area $\mathcal{R}$, then the normalized size of a hole H_i is $|H_i|/|\mathcal{R}|$. The *co-connectivity measure* CC is then defined as the sum of the normalized sizes of all holes:

$$\text{CC} = \sum_{i} |H_i|/|\mathcal{R}|$$

4.2 Simulation Setup and Methodology

In addition to PDT, we implement comparable aggregation algorithms discussed in Section 2. We implement all algorithms in Network Simulator-2 (NS-2) [24]. To provide a lower bound, we compute for each experiment an approximation of the optimal aggregation tree using the KMB algorithm. In our simulations, we collect the AVERAGE on a deployment of 750 nodes, placed randomly in a 75m x 75m grid. Each query collects data from a SN for 100 epochs. We utilize the NS-2 wireless communication infrastructure that simulates 914 MHz Lucent Wave LAN DSSS radio interface using the two ray ground reflection propagation model and IEEE 802.11 MAC layer (Chapters 16 & 18 of the NS-2 Manual [24]). Communication is performed using omni-directional antennas centered at each node, while the communication radius is fixed at 5m. The message payload is fixed at 72 bytes and we assume that every algorithm has the same payload for data transfers. Furthermore, we assume a lossless network with synchronized many-to-one aggregation, i.e., during in-network aggregation each internal node is perfectly synchronized with its children and after aggregation it always emits just one packet.

We use total data transmission (in MBs) as an indication of energy usage and hence as the basic metric of performance comparison. The amount of data transmission can be related to the energy expenditure by a simple function such as $\varepsilon = \sigma_s + \delta_s x$, where ε is the total amount of energy spent in sending a message with x bytes of content, and σ_s and δ_s represent the per-message and per-byte communication costs, respectively [9].

In order to systematically study the impact of varying participation levels and selective participation measures as defined in Section 4.1, we design our experiments according to the following questions:

– What is the impact of query selectivity and level of spatially correlated node participation on the performance of aggregation algorithms?
– What is the impact of the position of pockets and their dispersion in selective participation scenarios?
– What is the impact of the location of the sink on the performance of an aggregation algorithm in low node participation levels?
– What is the impact of (communication) holes on data collection?

4.3 Results

Impact of Query Selectivity. In two experiments, we investigate the impact of the query selectivity on four different aggregation schemes: PDT, multi-path (MP), static clustering (SC), and TAG. In the first experiment, Figure 2, we change the selectivity of an aggregate query and hence the number of nodes that participate in a query by 1% increments from 2% to 10% of the total nodes in the SN. The participating nodes are spatially clustered (see the deployment snapshot in Figure 2(a)). Figure 2(b) shows the mean value of the number of bytes transmitted by each algorithm at each participation level (the average of five runs is used to find the mean value). Figure 2(c) shows results from a similar experiment with participation levels ranging in discrete steps from 10% to 60%.

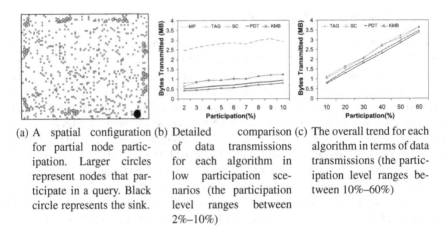

(a) A spatial configuration for partial node participation. Larger circles represent nodes that participate in a query. Black circle represents the sink.

(b) Detailed comparison of data transmissions for each algorithm in low participation scenarios (the participation level ranges between 2%–10%)

(c) The overall trend for each algorithm in terms of data transmissions (the participation level ranges between 10%–60%)

Fig. 2. The performance of aggregation techniques for varying levels of partial node participation

Figure 2(b) shows that for low node participation levels, PDT performs better than other aggregation schemes. For participation levels from 2% to 10% PDT is, on average, 41% more efficient than TAG and 37% more efficient than SC. In addition, PDT is just 21% less efficient than the approximated lower bound, where TAG and SC are 72% and 67% less efficient, respectively. This experiment also reveals that the energy consumption for MP in low participation scenarios is significantly higher than all other aggregation algorithms. MP requires at least 2.7 and 2.8 times as much data transmission as TAG and SC, respectively and 3.8 times more than PDT. Similarly, the trend in Figure 2(c) shows that PDT remains energy efficient even for high participation levels but its advantage decreases as the participation levels increase. At 10% participation, PDT requires 30% less data transmissions than TAG and 31% less than SC; however at the participation level of 60% this lead reduces to 4% and 5% for TAG and SC, respectively. The decrease in efficiency results from the fact that with the increase in node participation the benefit of spatial correlations diminishes. This effect can also be observed from the fact that at 60% participation level PDT is just 3% less efficient than the KMB lower bound. For high participation levels, MP is not shown on the figure to simplify the presentation.

Impact of Varying the Spatial Selectivity Index. This section describes a set of experiments that assess the performance of the PDT algorithm in various spatial layouts, characterized by different SSI values. A low SSI value represents a low dispersion scenario. We expect a better performance from PDT and other aggregation algorithms for selective queries with low SSI values. In this experiment, we achieve the effect of increasing SSI values by expanding the dispersion of pockets in the network, while the total number of pockets, the node participation level, and the sink position remains constant. Figure 3(a) and 3(b) show the deployment snapshots of two scenarios.

Figure 3(c) confirms the hypothesis that all algorithms perform better for lower SSI values (standard deviations are also shown in this chart). As the dispersion of the participating nodes increases, all algorithms have to spend more energy. For the analyzed

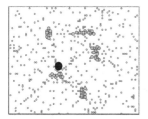 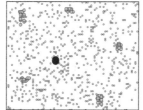 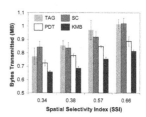

(a) A partial node participation scenario with an SSI of 0.34. Black circle represents the sink.

(b) A partial node participation scenario with an SSI of 0.57

(c) The impact of an increasing SSI on the performance of aggregation algorithms for a 10% node participation level. Standard deviations are shown.

Fig. 3. Various network configurations simulating an increase in the spatial selectivity index by increasing the pocket dispersion for a 10% node participation level

scenarios, TAG and SC transmit up to 31% and 22% more data for the highest SSI value. PDT also generates more data and shows an increase of 22% for the highest SSI value, however it remains 15% energy efficient than both TAG and SC.

Impact of the Location of the Sink. In this experiment we analyze the effect of sink position on PDT and other aggregation schemes. Figure 4 shows the performance of each algorithm in a deployment where the same query is issued from different sink positions. The chart shows that, for the given deployment, different sink positions affect the overall cost only modestly: between initial and final sink position data transmission rises by just 7% for both SC and TAG, while it rises to only 4% for PDT. The average change in cost from one scenario to next is 1%, 2% and 3% for PDT, SC, and TAG respectively.

The result is not surprising for the SC and PDT algorithms. In PDT the aggregation tree is mostly determined by the pocket locations while the impact of sink location is limited to the distance between the sink and the pockets closest to it. Similarly, SC always uses fixed paths to aggregate data inside each cluster and the impact of sink location is limited to the final phase where cluster heads have to route the aggregated data to the sink. The bulk of data transmission in both cases occurs inside pockets (or clusters) and as a result the impact of the position of sink is reduced. However, it is important to note that the behavior of TAG fluctuates with the sink location. Among the simulated scenarios, the second sink position is the best for TAG. At this position, the sink is located in a way that paths to distant pockets naturally emerge from the closer pockets, resulting in an increased number of participating nodes acting as intermediate nodes in the tree. In other scenarios, the misalignment of the root of the tree, sink, with the clusters increases the cost for TAG.

Spatial Layout and Communication Holes. Real SN deployments cannot stay fully connected in a regular grid structure although many routing and in-network aggregation

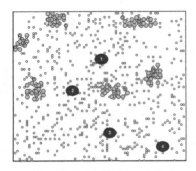 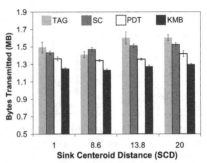

(a) The spatial distribution of the network configuration and the various positions of the sink

(b) The performance of the aggregation algorithms for different sink positions with an increasing distance to the global centroid position for a 20% node participation level

Fig. 4. A network configuration with different sink positions for a 20% node participation level

algorithms are commonly tested on such basic structures. Due to constrained communication capabilities, a network might be disconnected at certain places leaving gaps that we name as communication holes. In the context of in-network aggregation, if a given network suffer from communication gaps while still remaining connected via alternate communication routes, it is of interest to understand how the presence of holes effect the performance of an aggregation algorithm.

To investigate the effect of communication holes we simulate three different SN configurations. The configurations are shown in the deployment snapshots in Figure 5(a)–Figure 5(c), where the bordered regions represent communication holes. In each of the deployments, we set up a 10% pocketed participation scenario and Figure 5(d) shows the performance of each algorithm in these spatial layouts. In TAG, we see that a collection of communication holes can affect the formation of the aggregation tree in one of two ways. Firstly, the holes might break the most direct communication paths to pockets and hence the tree has to invariably take a longer route. This effect can be observed from the high cost of TAG in the first deployment (Figure 5(a)). However, a second more interesting scenario is where the presence of holes actually reduces the communication cost by restricting the tree into a set of corridors that naturally spans the pockets. The cost of TAG is reduced by 35% between the initial and final deployment.

SC is rather unaffected by the presence of holes where the change in its cost between the initial and final deployments is just 5%. PDT also performs almost unaffected by the presence of communication holes and shows a 10% reduction in data transmissions. PDT might suffer in cases where there is a communication gap between two neighboring pockets in the trajectory.

4.4 Discussion

With extensive experiments, we presented the advantages of the location based PDT algorithm for in-network aggregation over well-known in-network aggregation

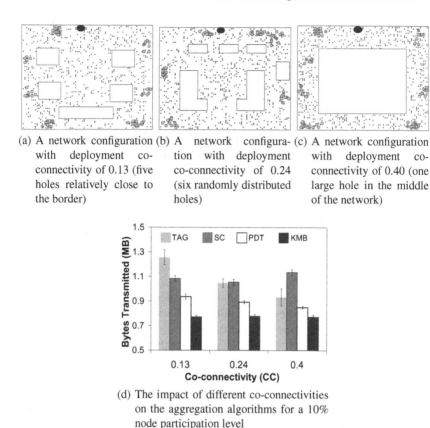

(a) A network configuration with deployment co-connectivity of 0.13 (five holes relatively close to the border)

(b) A network configuration with deployment co-connectivity of 0.24 (six randomly distributed holes)

(c) A network configuration with deployment co-connectivity of 0.40 (one large hole in the middle of the network)

(d) The impact of different co-connectivities on the aggregation algorithms for a 10% node participation level

Fig. 5. Various network configurations simulating deployments with different types of communication holes and a 10% node participation level

algorithms in different SN settings. Our results validate the hypothesis that a variable node participation scenario affects the performance of existing algorithms. In addition, the spatial features of the scenario do also have an effect on the performance.

The performance of an algorithm in processing selective queries can be presented as a function of the number of nodes the data collection paths utilize while collecting data from the participating nodes. The high cost of the tree-based scheme in highly selective queries can be explained by the random strategy used in the creation of an aggregation tree where no query specific optimization is considered during the tree construction process. We observe that for low node participation levels, the tree can improve its performance when the query sink is aligned with the data pockets in a way that paths to distant pockets naturally emerge from the pockets that are close to the sink. Similarly, a tree-based strategy shows considerable improvement in performance if the communication channels in the network are constrained by holes. However, in realistic settings, such cases may be rare and may not justify building a random tree for low participation scenarios. In contrast to such special cases, PDT always identifies constrained regions to grow a collection path in an efficient manner. Since the data collection is optimized to minimize the number of non-participating nodes en-route to data sources, PDT shows an overall reduction in data transmission even in high node participation scenarios. The experiments also show that the cost of PDT rises with increasing participation levels or decreasing spatial correlation levels, however for long running queries with many epochs, it is at least as good as the other well-known techniques.

As an interesting result, we observe that static clustering shows comparable results to the tree-based strategy even though it is not configured as dynamically as the tree-based strategy. Similar to the tree-based strategy, we also observe that static clustering performs better if the location and size of pockets correlate with that of the statically configured clusters and hence the data collection mechanism. The static clustering algorithm proposes to define the cluster size parameter depending upon the degree of spatial correlation in the network [17]. A major challenge in static clustering hence is determining the correct cluster size for queries with complex, selective, predicates.

In our experiments where the node participation levels are low, data transmission by the multi-path algorithm has greater costs than all the other schemes. This is an interesting observation since in exhaustive participation scenarios multi-path achieves the same number of messages per node as a tree [15]. Although the goal of the multi-path algorithm is to achieve accuracy in lossy environments rather than increasing efficiency of the aggregation strategy, the large cost of multi-path in low node participation scenarios may require adopting a hybrid method such as [15].

One important feature of the PDT algorithm is that there is a trade-off between the latency of data collection and the data transfer costs for a given query. In order to minimize the number of non-participating nodes in the aggregation process, PDT creates paths with possibly higher latencies. Since in a sensor network saving energy is the primary concern, reducing response time may not impact many query types. Thus, the latency is not analyzed as a separate parameter with our experiments.

5   Selectivity Under Changing Conditions

A central task for SN deployments is the monitoring of naturally occurring phenomena. Typical examples are the monitoring of wildlife paths in a natural reserve or the study of cattle movements in a farm. These application domains highlights the need for robust algorithms for selective query processing in SNs under continuously changing conditions. We analyze two types of change: (1) change of the physical phenomena that is monitored by the SN, and (2) change of the monitoring-network itself. The movement monitoring of wildlife in a natural reserve falls into the first category, while sensor failures are an example for the second category.

Changing physical phenomena pose a challenge for methods that are designed for processing selective queries. For a given selective query, the set of sensors that need to respond may be different at every sampling period. In fact, this behavior is expected as the main purpose of monitoring tasks is to observe and record changes in the physical phenomena in the first place. Thus, methods that are tailored for processing selective queries need to be able to cope with changing sets of participating nodes that respond to a query during the lifetime of that query.

We are currently adapting and experimenting with the PDT algorithm under realistic monitoring tasks as they occur with changing natural phenomena. The main insight for processing selective queries under changing conditions is that different observed states of the physical phenomena by a SN are in fact temporally as well as spatially correlated. Hence, the sensor readings do not change abruptly and randomly between two sampling periods but instead change occurs in relation to a source (or a set of sources), which leads to predictable patterns. A typical example is temperature measurements: the sensed temperature values are related to the Earth's revolution around the Sun. Similarly, cattle usually does not move randomly, but acts as a herd and moves within the constraints of the pasture, preferring certain places during day and night. Thus, we expect that finding efficient data collection paths under changing participation levels can be optimized using an approach that is based on PDT and includes predictive models. For low frequency sampling, the change between two periods could hide its behavior. However, as users are interested in capturing this behavior, the frequency of observations will be adjusted accordingly.

We envision two directions in adapting selectivity-aware algorithms such as PDT to monitor continuously changing physical phenomena: first, a reactive model that would only alter the data collection paths after a certain amount of change occurs in the sensor readings; second, a proactive model that predicts the changes and adjusts the collection path before the actual change occurs. A reactive algorithm might be easier to implement and cheaper to maintain but will be less responsive to changing physical phenomena. A proactive version, on the other hand, might be more difficult to maintain and develop, but may be more adaptive to the changing phenomena. It seems likely that the overall benefits of both strategies will differ for various application domains.

In addition to the change in the monitored physical phenomena, network parameters under which a SN operates can also change. We distinguish *uncontrolled* and *controlled* changes. An example for the first category is a production fault of some of the sensors in the network. This can occur independently from the deployment strategy and the physical conditions under which the SN operates. We assume that such situations rarely create a major disruption, and more importantly, do not specifically effect selective queries significantly. Recovery from such failures could be handled by the lower layers of the sensor node's system software. Nevertheless, PDT algorithm can be easily adapted for SNs where nodes can fail. The data collection paths can be tuned between sampling periods to bypass failing nodes efficiently.

Energy depletion rates are an example of a controlled change in a SN. Certain sensors may run out of energy faster than the others in a network, because they might be used more often in data collection paths. This type of change in a network is observable. Energy-aware algorithms that can adjust to these changes for routing already exist [25]. For selective query processing, techniques such as the PDT algorithm can also be adjusted to cope with similar situations. Location aggregation can be performed in tandem with the aggregation of the energy levels. Thus, data collection paths, per query, can be tailored for different energy levels of the nodes.

If the change in a SN is the result of a sustained use of certain set of sensor nodes, another alternative could be considered: redeployment of the sensor nodes. An interesting research direction is emerging with mobile sensor networks [26,27,28] where nodes can be redeployed for improving the quality of the data collected as well as for the efficiency of collection of data in tandem with improving the SN life expectancy.

6   Conclusions

Efficient processing of selective queries in SNs is crucial for effective sustained monitoring of physical phenomena. Existing data collection methods in SNs do not take an explicit position in processing selective queries and thus may not benefit from significant possible energy savings. In our work, we have presented the benefits of optimizing data collection paths and creating methods tailored for selective queries. We introduced the PDT algorithm, an in-network data collection method for long running selective queries. In extensive simulations, we showed that PDT is more energy efficient than other major aggregation techniques under various scenarios. We have also observed that the PDT algorithm computes a collection path to a well-known approximation of the global optimum, i.e., the minimal Steiner tree.

The efficient data collection problem in partial node participation scenarios can be modeled as a minimal Steiner tree problem. The PDT algorithm discovers pockets of participating nodes using purely local information about the network and approximates a minimum Steiner tree for data collection from these pockets. We show that this leads to significant energy savings in different node participation scenarios. In addition, we define spatial parameters to characterize a specific network deployment.

There are several research directions that we suggest as for future work. For a given query, the node participation can change over time. For example, a change in the physical conditions for the sensed environment can lead to changes in node participation. For a selective query this change can have an impact on multiple fronts, i.e., by increasing or decreasing the node participation levels, or by changing the distribution of participating nodes such as the emergence of new pockets or breakdown of old pockets. Therefore, it is important to introduce robust strategies for data collection to continuously adapt to these changes. On another front, changing network conditions can force the data collection algorithms to look for better ways of gathering data. Node failures, as a trivial case, can require algorithms, such as the PDT, to reconsider their choices in data collection paths. If the change in network conditions can be controlled, such as the energy depletion rates of different sensors, then adaptive selective query processing algorithms for controlled energy consumption should be considered. One particularly promising research direction is under scenarios when sensors have the capability of changing their location. In this case, increasing the quality of the data that is being collected in a partial node participation scenario while decreasing the data collection and node relocation costs is an interesting research topic.

References

1. Madden, S.R., Franklin, M.J., Hellerstein, J.M., Hong, W.: TinyDB: an acquisitional query processing system for sensor networks. ACM Trans. Database Syst. 30(1), 122–173 (2005)
2. Yao, Y., Gehrke, J.: Query processing for sensor networks. In: Proceedings of the Conference on Innovative Data Systems, pp. 233–244 (2003)

3. Madden, S., Franklin, M.J., Hellerstein, J.M., Hong, W.: TAG: a Tiny AGgregation service for ad-hoc sensor networks. SIGOPS Oper. Syst. Rev. 36(SI), 131–146 (2002)
4. Nath, S., Gibbons, P.B., Seshan, S., Anderson, Z.R.: Synopsis diffusion for robust aggregation in sensor networks. In: Proceedings of SenSys, pp. 250–262 (2004)
5. Madden, S., Franklin, M.J., Hellerstein, J.M., Hong, W.: The design of an acquisitional query processor for sensor networks. In: Proceedings of SIGMOD, pp. 491–502 (2003)
6. Oliveira, C.A.S., Pardalos, P.M.: A survey of combinatorial optimization problems in multicast routing. Comput. Oper. Res. 32(8), 1953–1981 (2005)
7. Kou, L., Markowsky, G., Berman, L.: A fast algorithm for Steiner trees. Acta Informatica 15, 141–145 (1981)
8. Takahashi, H., Matsuyama, A.: An approximate solution for the Steiner problem in graphs. Math Japonica 24, 573–577 (1980)
9. Silberstein, A., Braynard, R., Yang, J.: Constraint chaining: on energy-efficient continuous monitoring in sensor networks. In: Proceedings of SIGMOD, pp. 157–168 (2006)
10. Chou, J., Petrovic, D., Ramachandran, K.: A distributed and adaptive signal processing approach to reducing energy consumption in sensor networks. In: Proceedings of INFOCOM, vol. 2, pp. 1054–1062 (2003)
11. Bonfils, B.J., Bonnet, P.: Adaptive and Decentralized Operator Placement for In-Network Query Processing. In: Zhao, F., Guibas, L.J. (eds.) IPSN 2003. LNCS, vol. 2634, pp. 47–62. Springer, Heidelberg (2003)
12. Bawa, M., Gionis, A., Garcia-Molina, H., Motwani, R.: The price of validity in dynamic networks. In: Proceedings of the SIGMOD, pp. 515–526 (2004)
13. Considine, J., Li, F., Kollios, G., Byers, J.: Approximate aggregation techniques for sensor databases. In: Proceedings of ICDE, pp. 449–460 (2004)
14. Cardei, M., Wu, J.: Energy-efficient coverage problems in wireless ad hoc sensor networks. Computer Communications 29(4), 413–420 (2006)
15. Manjhi, A., Nath, S., Gibbons, P.B.: Tributaries and deltas: efficient and robust aggregation in sensor network streams. In: Proceedings of SIGMOD, pp. 287–298 (2005)
16. Chu, D., Deshpande, A., Hellerstein, J., Hong, W.: Approximate data collection in sensor networks using probabilistic models. In: Proceedings of ICDE, p. 48 (2006)
17. Pattem, S., Krishnamachari, B., Govindan, R.: The impact of spatial correlation on routing with compression in wireless sensor networks. In: Proceedings of IPSN, pp. 28–35 (2004)
18. Xu, Y., Heidemann, J., Estrin, D.: Geography-informed energy conservation for ad hoc routing. In: Proceedings of MobiCom, pp. 70–84 (2001)
19. Yoon, S., Shahabi, C.: Exploiting spatial correlation towards an energy efficient clustered aggregation technique (CAG). In: Proceedings of the ICC, pp. 82–98 (2005)
20. Gupta, H., Navda, V., Das, S.R., Chowdhary, V.: Efficient gathering of correlated data in sensor networks. In: Proceedings of MobiHoc, pp. 402–413 (2005)
21. Krishnamachari, B., Estrin, D., Wicker, S.B.: The impact of data aggregation in wireless sensor networks. In: Proceedings of ICDCSW, pp. 575–578 (2002)
22. Robins, G., Zelikovsky, A.: Improved Steiner tree approximation in graphs. In: Proceedings of SODA, pp. 770–779 (2000)
23. Doar, M., Leslie, I.M.: How bad is naive multicast routing? In: Proceedings of INFOCOM, pp. 82–89 (1993)
24. NS-2: The network simulator NS-2 documentation, http://www.isi.edu/nsnam/ns/ns-documentation.html
25. Yu, Y., Govindan, R., Estrin, D.: Geographical and energy aware routing: A recursive data dissemination protocol for wireless sensor networks. Technical Report TR-01-0023, University of California, Los Angeles, Computer Science Department (2001)

26. Somasundara, A.A., Jea, D.D., Estrin, D., Srivastava, M.B.: Controllably mobile infrastructure for low energy embedded networks. IEEE Transactions on Mobile Computing 5(8), 958–973 (2006)
27. Wang, G., Cao, G., Porta, T.F.L.: Movement-assisted sensor deployment. IEEE Transactions on Mobile Computing 5(6), 640–652 (2006)
28. Hull, B., Bychkovsky, V., Zhang, Y., Chen, K., Goraczko, M., Miu, A., Shih, E., Balakrishnan, H., Madden, S.: CarTel: a distributed mobile sensor computing system. In: Proceedings of SenSys, pp. 125–138 (2006)

Exploiting Spatio-temporal Correlations for Data Processing in Sensor Networks

Antonios Deligiannakis[1] and Yannis Kotidis[2]

[1] University of Athens
adeli@di.uoa.gr
[2] Athens University of Economics and Business
kotidis@aueb.gr

Abstract. Recent advances in microelectronics have made feasible the deployment of sensor networks for a variety of monitoring and surveillance tasks. In such tasks the state of the network is evaluated either at regular intervals at a base-station, which constitutes a centralized location where the data collected by the sensor nodes can be collected and processed, or continuously through the use of, potentially multiple, continuous queries. In order to increase the network lifetime, multiple techniques have been proposed in order to reduce the data transmitted in the network, since the data communication often constitutes the main source of energy drain in sensor networks. In this work we discuss several data reduction techniques that can be applied for energy-efficient query processing in sensor network applications. All of our proposed techniques seek to identify and take into account the characteristics of the collected data. Depending on the nature of the monitoring application at hand, the targeted data characteristics may range from simply monitoring the variance of a node's measurements to identifying spatio-temporal correlations amongst the values collected by the sensor nodes.

1 Introduction

Recent advances in wireless technologies and microelectronics have made feasible, both from a technological as well as an economical point of view, the deployment of densely distributed sensor networks [61]. Although today's sensor nodes have relatively small processing and storage capabilities, driven by the economy of scale, it is already observed that both are increasing at a rate similar to Moore's law.

In applications where sensors are powered by small batteries and replacing them is either too expensive or impossible (i.e., sensors thrown over a hostile environment), designing energy efficient protocols is essential to increase the lifetime of the sensor network. Since radio operation is by far the biggest factor of energy drain in sensor nodes [18], minimizing the number of transmissions is vital in data-centric applications. Even in the case when sensor nodes are attached to larger devices with ample power supply, reducing bandwidth consumption may still be important due to the wireless, multi-hop nature of communication and the short-range radios usually installed in the nodes.

S. Nittel, A. Labrinidis, and A. Stefanidis (Eds.): GSN 2006, LNCS 4540, pp. 45–65, 2008.
© Springer-Verlag Berlin Heidelberg 2008

Data-centric applications thus need to devise novel dissemination processes for minimizing the number of messages exchanged amongst the nodes. Nevertheless, in densely distributed sensor networks there is an abundance of information that can be collected. In order to minimize the volume of the transmitted data, we can apply two well known ideas *aggregation* and *approximation* in order to exploit spatio-temporal correlations in the readings obtained by the nodes in the network.

In-network aggregation is more suitable for exploratory, continuous queries that need to obtain a live estimate of some (aggregated) quantity. For example, sensors deployed in a metropolitan area can be used to obtain estimates on the number of observed vehicles. Temperature sensors in a warehouse can be used to keep track of the average and maximum temperature for each floor of the building. Often, aggregated readings over a large number of sensors nodes show little variance, providing a great opportunity for reducing the number of (re)transmissions by the nodes when individual measurements change only slightly (as in temperature readings) or changes in measurements of neighboring nodes effectively cancel out (as in vehicle tracking applications).

Approximation techniques, in the form of lossy data compression are more suitable for the collection of historical data through long-term queries. As an example, consider sensors dispersed in a wild forest, collecting meteorological measurements (such as pressure, humidity, temperature) for the purpose of obtaining a long term historical record and building models on the observed eco-system [2,42,67]. Each sensor generates a multi-valued data feed and often substantial compression can be achieved by exploiting natural correlations among these feeds (such as in case of pressure and humidity measurements). In such cases, sensor nodes are mostly "silent" (thus preserving energy) and periodically process and transmit large batches of their measurements to the monitoring station for further processing and archiving.

While the preferred method of data reduction, either by aggregation or by approximation, of the underlying measurements can be decided based on the application needs, there is a lot of room for optimization at the network level as well. Sensor networks are inherently redundant; a typical deployment uses a lot of redundant sensor nodes to cope with node or link failures [4]. Thus, extracting measurements from all nodes in the network for the purpose of answering a posed query may be both extremely expensive and unnecessary. In a data-centric network, nodes can coordinate with their neighbors and elect a small set of *representative nodes* among themselves, using a localized, data-driven bidding process [35]. These representative nodes constitute a *network snapshot* that can, in turn, answer posed queries while reducing substantially the energy consumption in the network. These nodes are also used as an alternative means of answering a posed query when nodes and network links fail, thus providing unambiguous data access to the applications.

This article proceeds as follows. Section 2 provides a brief overview of the characteristics of sensor nodes. Section 3 presents our Self-Based Regression (SBR) algorithm for the compression of historical measurements in sensor network applications, while Section 4 presents our framework for the approximate evaluation of continuous aggregate queries. Section 5 presents our techniques for creating a *network snapshot* that can be used to efficiently evaluate queries about the observed values of the sensor nodes, while Section 6 presents some related work in the area of sensor networks. Finally, Section 7 contains concluding remarks and future directions.

2   Characteristics of Sensor Nodes

Depending on the targeted application, sensor nodes with widely different characteristics may be used. Even though the processing and memory capabilities of sensor nodes are still limited, in recent years they have increased at a rate similar to Moore's law. On the other hand, the amount of energy stored in the batteries used in such nodes has exhibited a mere 2-3% annual growth. Since replacing the sensor batteries may be very expensive, and sometimes impossible due to their unattended deployment, unless the sensors are attached to and powered by a larger unit, designing energy-efficient protocols is essential to increase the lifetime of the sensor network.

The actual energy consumption by each sensor node depends on its current state. In general, each sensor node can be in one of the following states:

- *low-duty cycle*, where the sensor is in sleep mode and a minimal amount of energy is consumed.
- *idle listening*, where the sensor node is listening for possible data intended for it.
- *processing*, where the sensor node performs computation based on its obtained measurements and its received data.
- *receiving/transmitting*, where the sensor node either receives or transmits data or control messages.

The cost of processing can be significant but is generally much lower than the cost of transmission. For example, in the Berkeley MICA nodes sending one bit of data costs as much energy as 1,000 CPU instructions [41]. For long-distance radios, the transmission cost dominates the receiving and idle listening costs. For short-range radios, these costs are comparable. For instance, in the Berkeley MICA2 motes the power consumption ratio of transmitting/receiving at 433MHz with RF signal power of 1mW is 1.41:1 [64], while this ratio can become even larger than 3:1 for the same type of sensor when the radio transmission power is increased [54]. To increase the lifetime of the network, some common goals of sensor network applications are (in order of importance) to maximize the time when a node is in a low-duty cycle, to reduce the amount of transmitted and received data, and to reduce the idle listening time. We note here that reducing the size of the transmitted data results in multiple benefits, since this also corresponds to a reduction of not only control messages, but also leads to fewer message collisions and retransmissions. Moreover, nodes that refrain from transmitting messages may switch to the low-duty cycle mode faster, therefore further reducing their energy drain.

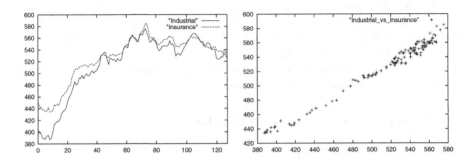

Fig. 1. Example of two correlated signals (Stock Market)

Fig. 2. XY scatter plot of Industrial (X axis) vs Insurance (Y axis)

3   A Lossy Compression Framework for Historical Measurements

Many real signals observed by the sensors, such as temperature, dew-point, pressure etc. are naturally correlated. The same is often true in other domains. For instance, stock market indexes or foreign currencies often exhibit strong correlations. In Figure 1 we plot the average Industrial and Insurance indexes from the New York stock market for 128 consecutive days. Both indexes show similar fluctuations, a clear sign of strong correlation. Figure 2 depicts a XY scatter plot of the same values. This plot is created by pairing values of the Industrial (X-coordinate) and Insurance (Y-coordinate) indexes of the same day and plotting these points in a two-dimensional plane. The strong correlation among these values makes most points lie on a straight line. This observation suggests the following compression scheme, inspired from regression theory. Assuming that the Industrial index (call it X) is given to us in a time-series of 128 values, we can approximate the other time-series (Insurance: Y) as $Y' = a * X + b$. The coefficients a and b are determined by the condition that the sum of the square residuals, or equivalently the L_2 error norm $||Y' - Y||_2$, is minimized. This is nothing more than standard linear regression. However, unlike previous methods, we will not attempt to approximate each time-series independently using regression. In Figure 1 we see that the series themselves are not linear, i.e., they would be poorly approximated with a linear model. Instead, we will use regression to approximate piece-wise correlations of each series to a base signal X that we will choose accordingly. In the example of Figure 2, the base signal can be the Industrial index (X) and the approximation of the Insurance index will be just two values (a, b). In practice the base signal will be much smaller than the complete time series, since it only needs to capture the "important" trends of the target signal Y. For instance, in case Y is periodic, a sample of the period would suffice.

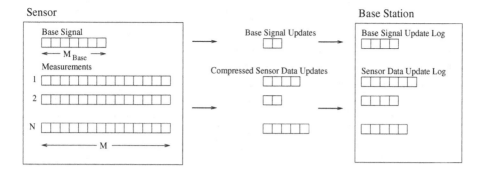

Fig. 3. Transfer of approximate data values and of the base signal from each sensor to the base station

3.1   The SBR Framework

In the general case, each sensor monitors N distinct quantities $\boldsymbol{Y_i}$, $1 \leq i \leq N$. Without loss of generality we assume that measurements are sampled with the same rate. When enough data is collected (for instance, when the sensor memory buffers become full), the latest NxM values are processed and each row i (of length M) is approximated by a much smaller set of B_i values, i.e. $B_i \ll M$. The resulting "compressed" representation, of total size equal to $\sum_{i=1}^{N} B_i$, is then transmitted to the base station. The base station maintains the data in this compact representation by appending the latest "chunk" to a log file. A separate file exists for each sensor that is in contact with the base station. This process is illustrated in Figure 3. Each sensor allocates a small amount of memory of size M_{base} for what we call the *base signal*. This is a compact ordered collection of values of prominent features that we extract from the recorded values and are used as a base reference in the approximate representation that is transmitted to the base station. The data values that the sensor transmits to the base station are encoded using the in-memory values of the base signal at the time of the transmission. The base signal may be updated at each transmission to ensure that it will be able to capture newly observed data features and that the obtained approximation will be of good quality. When such updates occur, they are transmitted along with the data values and appended in a special log file that is unique for each sensor.

The Self-Based Regression algorithm (SBR) breaks the data intervals $\boldsymbol{Y_i}$ into smaller data segments

$$I_i[k..l] = (Y_i[k], \ldots Y_i[l])$$

and then pairs each one to an interval of the base signal of equal length. As discussed below, the base signal is simply the concatenation of several intervals of the same length W extracted from the data. The data interval I_i is shifted over the base signal and at each position s we compute the regression parameters for the approximation

$$\hat{I}_i[j] = a \times X[s+j-k] + b, k \le j \le l$$

and retain the shift value $s = s^*$ that minimizes the sum-squared error of the approximation. The algorithm starts with a single data interval for each row of the collected data (Y_i). In each iteration, the interval with the largest error in the approximation is selected and divided in two halves. The compressed representation of a data interval $I_i[k..l]$ consists of four values: the shift position s^* that minimizes the error of the approximation, the two regression parameters a,b and the start of the data interval k in Y_i. The base station will sort the intervals based on their start position and, thus, there is no need to transmit their ending position. Given a target budget B (size of compressed data) we can use at most $B/4$ intervals using this representation.

3.2   Base Signal Construction

We can think of the base signal as a dictionary of features used to describe the data values. The richer the pool of features we store in the base signal the better the approximation. On the other hand, these features have to be (i) kept in the memory of the sensor to be used as a reference by the data-reduction algorithm and (ii) sent to the base station in order for it to be able to reconstruct the values. Thus, for a target bandwidth constraint B (number of values that can be transmitted), performing more insert and update operations on the base signal implies less bandwidth remaining for approximating the data values, and, thus, fewer data intervals that can be obtained from the recursive process described above.

We can avoid the need of transmitting the base signal by agreeing a-priori on a set of functions that will be used in the regression process. For instance, a set of cosine functions (as in the Distinct Cosine Transform) can be used for constructing a "virtual" base signal that does not need to be communicated. Similarly, using the identity function $X[i] = i$ reduces the compression algorithm to standard linear regression of each data interval. However, such an approach makes assumptions that may not hold for the data at hand. In [11] we have proposed a process for generating the base signal from the data values. The key idea is to break the measurements into intervals of the same length W. Each interval (termed *candidate base interval*) is assigned a score based on the reduction in the error of the approximation obtained by adding the interval to the base signal. Using a greedy algorithm we can select the top-few candidate intervals, up to the amount of available memory M_{base}. Then a binary-search process is used to eventually decide how many of those candidate intervals need to be retained.

The search space is illustrated in Figure 4 for three real data sets, discussed in [11]. The figure plots the error of only the initial transmission as the size of the base signal is varied, manually, from 1 to 30 intervals. We further show the selection of the binary-search process. For presentation purposes, the errors for each data set have been divided by the error of the approximation when using just one interval. We notice that initially, by adding more candidate intervals

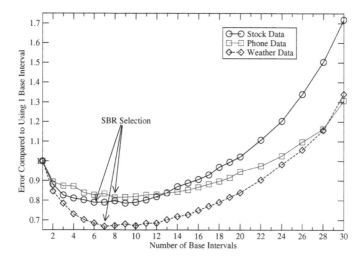

Fig. 4. SSE error vs base signal size

to the base signal the error of the approximation is reduced. However, after a point, adding more intervals that need to be transmitted to the base station leaves insufficient budget for the recursive process that splits the data, and thus, the error of the approximation is eventually increased.

3.3 Analysis and Evaluation

For a data set containing $n = N \times M$ measurements to approximate the complete SBR algorithm takes $O(n^{1.5})$ time and requires linear space, while its running time scales linearly to the size of both the transmitted data and the base signal. The algorithm is invoked periodically, when enough data has been collected. Moreover, our work [11] has demonstrated that, after the initial transmissions, the base signal is already of good quality and few intervals are inserted in it. This provides us with the choice not to update the base signal in many subsequent invocations, thus reducing the algorithm's running time, in these cases, to only a linear dependency on n.

To illustrate the accuracy achieved by the SBR algorithm against standard approximation techniques such as Wavelets, Histograms and the Discreet Cosine Transform (DCT), we used in [11], among other data sets, a weather data set that contains the air temperature, dewpoint temperature, wind speed, wind peak, solar irradiance and relative humidity weather measurements obtained from a station in the university of Washington, and for the year 2002. For this data set we selected the first 40,960 records and then split the data measurements of each signal into ten files of 4,096 values each, in order to simulate multiple transmissions. We then varied the compression ratio (size of the transmitted data over the data size n) from 5% to 50% and present in Table 1 the total sum

Table 1. Total SSE Varying the Compression Ratio for the Weather Data Set

Compression Ratio	Weather Data (245,760 total values)			
	SBR	Wavelets	DCT	Histograms
5%	317,238	519,303	8,703,192	7,661,293
10%	103,633	200,501	4,923,294	3,375,518
15%	54,219	125,449	3,515,698	2,219,533
20%	30,946	87,118	2,643,229	1,471,421
25%	18,600	63,105	2,198,455	946,735
30%	11,558	46,833	1,598,451	594,644
35%	7,161	35,275	1,366,211	410,208
40%	4,603	26,824	1,112,117	288,127
45%	2,964	20,502	905,422	236,947
50%	1,861	15,762	768,568	160,079

squared error achieved by all techniques. In all cases, SBR produces significantly more accurate results than the other approximations.

3.4 Extensions

In [14] we present extensions to the basic SBR scheme that we describe here. These extensions allow the nodes to organize in groups based on an adaptation of the HEED protocol [65] and elect within each group a group leader that instruments the execution of each SBR instance in the nodes of its group and also handles the final transmission of the compressed data to the base station. This form of localized processing allows the nodes to exploit spatial correlations and results in a reduction of the error of the approximation by at least an order of magnitude compared to the case when nodes individually compress and transmit their data.

4   Approximate In-Network Data Aggregation

The data aggregation process in sensor networks consists of several steps. First, the posed query is disseminated through the network, in search of nodes collecting data relevant to the query. Each sensor then selects one of the nodes through which it received the announcement as its *parent node*. The resulting structure is often referred to as the *aggregation tree*. Non-leaf nodes of that tree aggregate the values of their children before transmitting the aggregate result to their parents. In [40], after the aggregation tree has been created, the nodes carefully schedule the periods when they transmit and receive data. The idea is for a parent node to be listening for values from its child nodes within specific intervals of each *epoch* and then transmit upwards a single partial aggregate for the whole subtree.

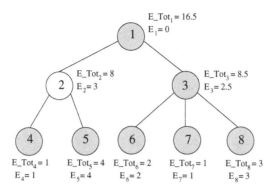

Fig. 5. Error Filters on Aggregation Tree

In order to limit the number of transmitted messages and, thus, the energy consumption in sensor networks during the computation of continuous aggregate queries, our algorithms install error filters on the nodes of the aggregation tree. Each node N_i transmits the partial aggregate that it computes at each epoch for its subtree only if this value deviates by more than the maximum error E_i of the node's filter from the last transmitted partial aggregate. This method allows nodes to refrain from transmitting messages about small changes in the value of their partial aggregate. The E_i values determine the maximum deviation E_Tot_i of the reported from the true aggregate value at each node. For example, for the SUM aggregate, this deviation at the monitoring node is upper bounded (ignoring message losses) by: $\sum_i E_i$.

A sample aggregation tree is depicted in Figure 5. As our work [12] has demonstrated, allowing a small error in the reported aggregate can lead to dramatic bandwidth (and thus energy) savings. The challenge is, of course, given the maximum error tolerated by the monitoring node, to calculate and periodically adjust the node filters in order to minimize the bandwidth consumption.

4.1 Algorithmic Challenges

When designing adaptive algorithms for in-network data aggregation in sensor networks, one has to keep in mind several challenges/goals. First, communicating *individual* node statistics is very expensive, since this information cannot be aggregated inside the tree, and may outweigh the benefits of approximate data aggregation, namely the reduction in the size of the transmitted data. Thus, our algorithm should not try to estimate the number of messages generated by each node's transmissions, since this depends on where this message is aggregated with messages from other nodes. Second, the error budget should be distributed to the nodes that are expected to reduce their bandwidth consumption the most by such a process. This benefit depends on neither the magnitude of the partial aggregate values of the node nor the node's number of transmissions over a period, but on the magnitude of the changes on the calculated partial aggregate.

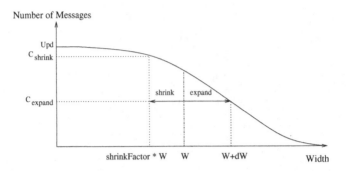

Fig. 6. Potential Gain of a Node

Isolating nodes with large variance on their partial aggregate and redistributing their error to other nodes is crucial for the effectiveness of our algorithm [12]. Finally, in the case of nodes where the transmitted differences from their children often result in small changes on the partial aggregate value, our algorithm should be able to identify this fact. We deal with this latter challenge by applying the error filters on the calculated partial aggregate values and not on each node's individual measurements. For the first two challenges, we collect a set of easily computed and composable statistics at each node. These statistics are used for the periodic adjustment of the error filters.

4.2 Algorithm Overview

Every Upd epochs all nodes shrink the widths of their filters by a shrinking factor $0 < shrinkFactor < 1$. After this process, the monitoring node has an error budget of size $E\_Global \times (1 - shrinkFactor)$, where E_Global is the maximum error of the application, that it can redistribute recursively to the nodes of the network. Each node, between two consecutive update periods, calculates its *potential gain* as follows: At each epoch the node keeps track of the number of transmissions C_{shrink} that it would have performed with the default (smaller) filter at the next update period of width $shrinkFactor \times W_i$, where $W_i = 2 \times E_i$. The node also calculates the corresponding number of transmissions C_{expand} with a larger filter of width $W_i + dW$ and sets its potential gain to $Gain_i = C_{shrink} - C_{expand}$. This process is illustrated in Figure 6. The *cumulative gain* of a node's subtree is then calculated as the sum of the cumulative gains of the node's children and the node's potential gain. This requires only the transmission of the cumulative gains (a single value for each node) at the last epoch before the new update period. The available error budget is then distributed top-down proportionally, at each node, to each subtree's cumulative gain. In this process, nodes that exhibit large variance in their partial aggregate values will exhibit small potential gains and, thus, their error will gradually shrink and be redistributed to nodes that will benefit from an increase in their error filter.

We note here that the dual problem of minimizing the application maximum error given a bandwidth or energy constraint is also very interesting. This

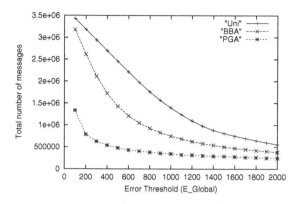

Fig. 7. Reduction in messages for synthetic data set

problem is discussed in [13] and is more complicated, though, because the controlled quantity at each node (the width of its error filter) is different from the monitored quantity (the bandwidth/energy consumption) and the bandwidth needs to be carefully computed, monitored and then disseminated amongst the sensor nodes.

4.3   Experimental Evaluation

We used a balanced aggregation tree with 5 levels and a fan-out of 4 (341 nodes overall), where all the nodes collected measurements relevant to the query. In the first experiment, the measurements of all the nodes followed a random walk pattern, each with a randomly assigned step size in the range $(0 \ldots 2]$. To capture the scenario where nodes update their measurements with different frequencies, 20% of the nodes update their measurements at each epoch, while the remaining sensors make a random step with a fixed probability of 1% during an epoch. In Figure 7 we plot the total number of messages in the network (y-axis) for 40,000 epochs when varying the error constraint E_Global from 100 to 2,000 (8% is terms of relative error). Depending on E_Global, our PGA (Potential Gains Adjustment) algorithm results in up to 4.8 times fewer messages than an adaptation of the algorithm proposed by Olston in [45], that we termed BBA, and up to 6.4 times fewer messages than a uniform allocation policy (Uni) where the error is partitioned uniformly amongst all nodes. These differences arise from the ability of PGA to place, judiciously, filters on passive intermediate sensor nodes and exploit negative correlations on their subtree based on the computed potential gains. Algorithm BBA may also place filters on the intermediate nodes (when the residual mode is used) but the selection of the widths of the filters based on the burden scores of the nodes was typically not especially successful in our experiments.

5   Design of Data-Centric Sensor Networks

Sensor networks are inherently dynamic. Such networks must adapt to a wide variety of challenges imposed by the uncontrolled environment in which they operate. As nodes become cheaper to manufacture and operate, one way of addressing the challenges imposed on unattended networks is redundancy [4]. Redundant nodes ensure network coverage in regions with non-uniform communication density due to environmental dynamics. Redundancy further increases the amount of data that can be mined in large-scale networks.

Writing data-driven applications in such a dynamic environment can be daunting. The major challenge is to design localized algorithms that will perform most of the processing in the network itself in order to reduce traffic and, thus, preserve energy. Instead of designing database applications that need to hassle with low-level networking details, we envision the use of data-centric networks that allow transparent access to the collected measurements in a unified way. For instance, when queried nodes fail, the network should self-configure to use redundant stand-by nodes as in [18], under the condition that the new nodes contain fairly similar measurements, where similarity needs to be quantified in an application-meaningful way [35]. This can be achieved using a localized mode of operation in which nodes coordinate with their neighbors and elect a small set of *representative nodes* among themselves. Such a set of representatives, termed *network snapshot* [35], has many advantages.

- The location and measurements of the representative nodes provide a picture of the value distribution in the network. By choosing an error metric (such as sum-squared or relative error) and using different threshold values to express similarity amongst the sensor node measurements we can obtain different snapshots of the network at different resolutions, depending on the error threshold used.
- The network snapshot can be used for answering user queries in a more energy-efficient way. The data reduction techniques that we discussed in Sections 3 and 4 aim at reducing the flow of data in the network by either suppressing update messages or compressing long data streams. The network snapshot is an orthogonal optimization that can further reduce energy drain during query processing by reducing the number of nodes that need to respond to user queries. When a user query can tolerate a small error in the computation, the network can use the representative nodes and compile a quick answer from only a fraction of the nodes that a normal query execution would require. We call such queries *snapshot queries*.
- An elected representative node can take over for another node in the vicinity that may have failed or is temporarily out of reach. Because selection of representatives is quantitative this allows for a more accurate computation than when representatives are selected based only on proximity.
- A localized computation of representative nodes can react quickly to changes in the network. For instance, nodes (including the representatives) may fail at random. It is important that the network can self-heal in the case of

node-failures or some other catastrophic events. In a data-driven mode of operation, we are also interested in changes in the behavior of a node. In such case the network should re-configure and revise the selected representatives, when necessary. What is important is that, as we demonstrate in [35], these computations can be performed in the network with only a small number (up to six) of exchanged messages among the nodes.

5.1   Snapshot Overview

A sensor node N_i maintains a data model for capturing the distribution of values of the measurements of its neighbors. This is achieved by snooping (with a small probability) values broadcast by its neighbor node N_j in response to a query or, by using periodic announcements sent by N_j. One may devise different data models, with varying degrees of complexity, for this process. In [11,35] we have proposed modeling the correlations amongst the measurements of the nodes using linear regression. Regression models are simple both in terms of space and time complexity. As is demonstrated in [35], our algorithms can operate when the available memory for storing these models in the sensor is as small as a few bytes. Furthermore, by modeling the correlations amongst the values of the nodes we avoid making assumptions on the distribution of the data values collected by the sensors that may not hold in practice. The only assumption made is that values of neighboring nodes are to some extent correlated. This is true for many applications of interest like collection of meteorological data, acoustic data etc, as discussed in Section 3.

Using the data model it maintains, sensor node N_i provides an estimate $\hat{x}_j$ of the measurement x_j of its neighbor N_j. Given an error metric $d()$ and a threshold value T, node N_i can *represent* node N_j if $d(x_j, \hat{x}_j) \leq T$. Function $d()$ is provided by the application. Some common choices include (i) relative error: $d(x_j, \hat{x}_j) = \frac{|x_j - \hat{x}_j|}{\max(s, |x_j|)}$, where $s > 0$ is a sanity bound for the case $x_j=0$, (ii) absolute error: $d(x_j, \hat{x}_j) = |x_j - \hat{x}_j|$ and (iii) sum-squared error: $d(x_j, \hat{x}_j) = (x_j - \hat{x}_j)^2$.

Through a localized election process (see [35] for details) the nodes in the network pick a set of representative nodes of size n_1. Depending on the threshold value T, the error metric and the actual measurements on the sensors, n_1 can be significantly smaller than the number of nodes in the network. An example of this process is demonstrated in Figure 8 where the representatives for a simulated network of 100 nodes are shown. Dark nodes in the Figure are representative nodes. There are lines drawn from a node N_i to a node N_j that N_i represents. Nodes with no lines attached to them represent themselves (the default choice).

An aggregate computation like SUM can be handled by the representative nodes that will in-turn provide estimates on the nodes N_j they represent using their models. Another scenario is to use the representative of a node on an aggregate or direct query, when that node is out-of-reach because of some unexpected technical problem or due to severe energy constraints. Thus, query processing can take advantage of the unambiguous data access provided by the network. Of course, one can ignore the layer of representatives and access the sensors directly,

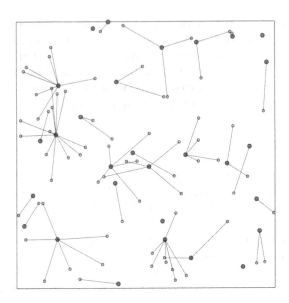

Fig. 8. Example of Network Snapshot

at the penalty of (i) draining more energy, since a lot more nodes will have be to accessed for the same query and (ii) having to handle *within the application* node failures, redundancy etc.

The selection of representatives is not static but is being revised overtime in an adaptive fashion. An obvious cause is the failure of a representative node. In other cases, the model built by N_i to describe x_j might get outdated or fail, due to some unexpected change in the data distribution. In either case, the network will self-heal using the following simple protocol. Node N_j periodically sends a heart-beat message to its representative N_i including its current measurement. If N_i does not respond, or its estimate $\hat{x}_j$ for x_j is not accurate ($d(x_j, \hat{x}_j) > T$) then N_j initiates a local re-evaluation process inviting candidate representatives from its neighborhood, leading to the selection of a new representative node (that may be itself). This heart-beat message is also used by N_i to fine-tune its model of N_j.

Under an unreliable communication protocol it is possible that this process may lead to *spurious* representatives. For instance node N_i may never hear the messages sent by node N_j due to an obstacle in their direct path. It may thus assume that it still represents node N_j while the network has elected another representative. This can be detected and corrected by having time-stamps describing the time that a node N_i was elected as the representative of N_j and using the latest representative based on these time-stamps. In TinyOS nodes have an external clock that is used for synchronization with their neighbors [40]. In lack of properly synchronized clocks among the sensor nodes, one can use a global counter like the epoch-id of a continuous query. This filtering and self-correction is performed by the network, transparently from the application.

5.2   Examples of Snapshot Queries

Recent proposals [40,63] have suggested the use of SQL for data acquisition and processing. The obvious advantage of using a declarative language is greater flexibility over hand-coded programs that are pre-installed at the sensor nodes [41]. In addition embedded database systems like TinyDB can provide energy-based query optimization because of their tight integration with the node's operations.

Basic queries in sensor networks consist of a SELECT-FROM-WHERE clause (supporting joins and aggregation). For instance, in our running example of collecting weather data a typical query may be

```
SELECT loc, temperature
FROM sensors
WHERE loc in SHOUTH_EAST_QUANDRANT
SAMPLE INTERVAL 1sec for 5min
USE SNAPSHOT
```

This is an example of a drill-through query, sampling temperature readings every 1 second and lasting 5 minutes. For this example we assume that each node has a unique location-id loc and that nodes are location-aware, being able to process the spatial filter "in SHOUTH_EAST_QUANDRANT". Location can be obtained using inexpensive GPS receivers embedded in the nodes, or by proper coordination among the nodes, using techniques like those proposed in [50,52]. Location is a significant attribute of a node in an unattended system. For many applications like habitat monitoring, spatial filters may be the most common predicate.

The new extension presented in the query above is the USE SNAPSHOT clause, denoting that the query may be answered by the representative set of nodes. An example of an aggregate query that is using the snapshot for computing the average and maximum temperature readings in the same area is given below

```
SELECT avg(temperature), max(temperature)
FROM sensors
WHERE loc in SHOUTH_EAST_QUANDRANT
SAMPLE INTERVAL 1sec for 5min
USE SNAPSHOT
```

5.3   Evidence of Savings During Snapshot Queries

We used a simulated network of 100 sensor nodes, randomly placed in a $[0\ldots 1) \times [0\ldots 1)$ two-dimensional area. For each node, we generated values following a random walk pattern, each with a randomly assigned step size in the range $(0\ldots 1]$. The initial value of each node was chosen uniformly in range $[0\ldots 1000)$. We then randomly partitioned the nodes into K classes. Nodes belonging to the same class i were making a random step (upwards or downwards) with the same

Table 2. Reduction in number of nodes participating in a spatial snapshot query

	$K=1$		$K=100$	
	Transmission Range		Transmission Range	
Query Range	0.2	0.7	0.2	0.7
1%	11%	29%	3%	7%
10%	38%	77%	16%	24%
50%	52%	91%	23%	49%

probability $P_{move}[i]$. These probabilities were chosen uniformly in range $[0.2\ldots 1]$ (we excluded values less than 0.2 to make data more volatile).

We tested aggregate queries over random parts of the network. For each query a *sink* node was chosen randomly. Then, using the flooding mechanism described in [40] an aggregation tree was formed, rooted at the sink node. The sensor nodes N_i whose measures were aggregated using that tree, were chosen using spatial predicate "$location_i$ in $[x - \frac{W}{2}, x + \frac{W}{2}] \times [y - \frac{W}{2}, y + \frac{W}{2}]$", where (x,y) is a random point in the $[0\ldots 1) \times [0\ldots 1)$ plane.

We created a random set of 200 such queries and executed each query in the set twice: once as a regular query and once as a snapshot query. We counted the number of nodes participated in each execution, denoted as $N_{regular}$ and $N_{snapshot}$ respectively. In Table 2 we show the savings $\frac{N_{regular}-N_{snapshot}}{N_{regular}}$ provided on the average by the snapshot queries (the error threshold was one). We note that when snapshot queries are used, a non-representative node may still be used for routing the aggregate and this is included in the numbers shown. We made two runs, one with a single class and another when each node was on a class of its own ($K=100$). We varied the size W of the range queries as shown in the table. We further tested two transmission ranges for the nodes. The shorter transmission range results in more representatives and taller aggregation trees, as more hops are required to reach the sink node. In Table 2 we can see that snapshot queries provide substantial savings in terms of the number of nodes participating in a query, especially on large spatial queries. For all runs, the aggregation tree was created using the vanilla method of [10,40]. One can modify the protocol to favor (when applicable) representative nodes for routing the messages. This will result in further reduction in the number of sensor nodes used during snapshot queries than those presented in Table 2.

6   Related Work

In recent years, there has been a significant body of work in the area of sensor networks. For instance, the networking aspects of wireless sensor nodes is a topic that has intrigued the networking community. Because of the unattended operation of sensor networks, nodes must be able to co-operate to perform the task at hand. Some of the most important topics addressed include network self-configuration [4,35,65], discovery [18,27] and computation of energy-efficient data routing paths [6,28,39,40,55,57,65].

In the database community there are ongoing projects for infusing database primitives in the operations of these networks. For instance, TinyDB [41] and Cougar [63] have suggested the use of SQL for data acquisition and processing. The obvious advantage of using a declarative language is greater flexibility over hand-coded programs that are pre-installed at the sensor nodes [41]. In-network data aggregation is another topic that has created a flurry of proposals [12,16,19,30,40,53,63]. The main idea is to build an aggregation tree, which partial results will follow. With proper synchronization [40], non-leaf nodes of the tree aggregate the values of their children before transmitting a single aggregate result to their parents, thus substantially reducing the flow of messages in the network. Alternative, gossip-based techniques have also been investigated in [3,33]. In [10,35] the authors have also looked at the problems of packet loss and node failures during data processing. Recently, proposals for combining data modeling with data acquisition in order to help reduce the cost of query processing have been suggested [17,35,37]. For example, [17,37] build probabilistic models of the observed data and then use these models to probe the sensors for their measurements in a limited amount of epochs, depending on the confidence of the constructed model. Distributed storage management is another topic that brings together the networking and database communities [16,21,51].

Many of these fundamental techniques have been devised to support event-based monitoring applications. For example, in animal tracking, an event such as the presence of an animal can be determined by matching the sensor readings to stored patterns [31]. The authors of [62] propose an event detection mechanism based on matching the contour maps of in-network sensory data distributions. In [47], kernel-based techniques are used to detect abnormal behavior in sensor readings. In [26] the authors describe the implementation of a real system based on Mica2 motes for surveillance of moving vehicles. In [36] a framework for computing user-defined events that are in proximity is presented.

Query processing in sensor networks has some connection with the work on continuous queries in data streams [7,29,44,59,66]. The work of [46] studies the trade-off between precision and performance when querying replicated, cached data. In [45] the users register continuous queries with strict precision constraints at a central *stream processor*, which, in turn installs filters at the remote data sources. These filters adapt to changes in the streams to minimize update rates. Online algorithms for minimizing the update cost while the query can be answered within an error bound are presented in [34]. The authors of [9] study a probabilistic query evaluation method that places appropriate confidence in the query answer to quantify the *uncertainty* of the recorded data values.

There is a vast related literature on approximate processing techniques. The AQUA project explored sampling-based techniques for building *synopses* and using them to provide approximate answers at a fraction of the time that a real answer would require [22]. Histograms are used by query optimizers to estimate the selectivity of queries, and recently in tools for providing fast approximate answers to queries [23,24,32,48,49,58]. Wavelets are a mathematical tool for the hierarchical decomposition of functions, with applications in image and signal processing [56]. More recently, Wavelets have been applied successfully in answering range-sum aggregate queries over data cubes [60], in selectivity estimation [43] and in approximate query processing [5,15,20,25]. The Discrete Cosine Transform (DCT) [1] constitutes the basis of the *mpeg* encoding algorithm and has also been used to construct compressed multidimensional histograms [38]. Linear regression has been recently used in [8] for on-line multidimensional analysis of data streams.

7   Conclusions and Future Directions

We have described several techniques for the reduction of the transmitted data in several sensor network applications, ranging from the communication of historical measurements to answering approximate aggregate continuous and snapshot queries. While these techniques aim to prolong the lifetime of the network, there are several issues that need to be additionally addressed. Little work has been done on the optimization of multiple concurrent continuous queries over sensor networks. The work of Olston et al. in [45] may provide some helpful solutions in this area. Moreover, in the presence of nodes with different transmission frequencies, as in the case of approximate aggregate query processing, several communication and synchronization algorithms may need to be revisited [63]. For example, the selection of the aggregation tree is often performed by assuming equal frequency of transmissions by all nodes. However, it might be more beneficial to prevent nodes that exhibit large variance in their measurements from appearing in lower levels of the tree, since such nodes often trigger transmissions on their ancestors as well. Such optimizations may lead to even larger energy savings.

References

1. Ahmed, N., Natarakan, T., Rao, K.R.: Discrete cosine transform. IEEE Trans. on Computers C-23 (1974)
2. Ailamaki, A., Faloutsos, C., Fischbeck, P.S., Small, M.J., Van Briesen, J.: An environmental sensor network to determine drinking water quality and security. SIGMOD Record 32(4), 47–52 (2003)
3. Bawa, M., Garcia-Molina, H., Gionis, A., Motwani, R.: Estimating Aggregates on a Peer-to-Peer Network. Technical report, Stanford (2003)
4. Cerpa, A., Estrin, D.: ASCENT: Adaptive Self-Configuring sEnsor Network Topologies. In: INFOCOM (2002)
5. Chakrabarti, K., Garofalakis, M., Rastogi, R., Shim, K.: Approximate Query Processing Using Wavelets. In: Proceedings of the 26th VLDB Conference (2000)
6. Chang, J.-H., Tassiulas, L.: Energy Conserving Routing in Wireless Ad-hoc Networks. In: INFOCOM (2000)
7. Chen, J., Dewitt, D.J., Tian, F., Wang, Y.: NiagaraCQ: A Scalable Continuous Query System for Internet Databases. In: Proceedings of ACM SIGMOD Conference (2000)

8. Chen, Y., Dong, G., Han, J., Wah, B.W., Wang, J.: Multi-Dimensional Regression Analysis of Time-Series Data Streams. In: Proceedings of VLDB (2002)
9. Cheng, R., Kalashnikov, D.V., Prabhakar, S.: Evaluating Probabilistic Queries over Imprecise Data. In: Proceedings of ACM SIGMOD Conference (2003)
10. Considine, J., Li, F., Kollios, G., Byers, J.: Approximate Aggregation Techniques for Sensor Databases. In: ICDE (2004)
11. Deligiannakis, A., Kotidis, Y., Roussopoulos, N.: Compressing Historical Information in Sensor Networks. In: Proceedings of ACM SIGMOD Conference (2004)
12. Deligiannakis, A., Kotidis, Y., Roussopoulos, N.: Hierarchical in-Network Data Aggregation with Quality Guarantees. In: Lindner, W., Mesiti, M., Türker, C., Tzitzikas, Y., Vakali, A.I. (eds.) EDBT 2004. LNCS, vol. 3268. Springer, Heidelberg (2004)
13. Deligiannakis, A., Kotidis, Y., Roussopoulos, N.: Bandwidth Constrained Queries in Sensor Networks. The VLDB Journal (2007)
14. Deligiannakis, A., Kotidis, Y., Roussopoulos, N.: Dissemination of Compressed Historical Information in Sensor Networks. The VLDB Journal (2007)
15. Deligiannakis, A., Roussopoulos, N.: Extended Wavelets for Multiple Measures. In: Proceedings of SIGMOD Conference, pp. 229–240 (2003)
16. Demers, A., Gehrke, J., Rajaraman, R., Trigoni, N., Yao, Y.: The Cougar Project: A Work In Progress Report. SIGMOD Record 32(4), 53–59 (2003)
17. Deshpande, A., Guestrin, C., Madden, S., Hellerstein, J.M., Hong, W.: Model-Driven Data Acquisition in Sensor Networks. In: Proceedings of VLDB (2004)
18. Estrin, D., Govindan, R., Heidermann, J., Kumar, S.: Next Century Challenges: Scalable Coordination in Sensor Networks. In: MobiCOM (1999)
19. Ganesan, D., Estrin, D., Heidermann, J.: DIMENSIONS: Why do we need a new Data Handling architecture for Sensor Networks? In: HotNets-I (2002)
20. Garofalakis, M., Gibbons, P.B.: Probabilistic Wavelet Synopses. ACM Trans. Database Syst. 29(1), 43–90 (2004)
21. Ghose, A., Grossklags, J., Chuang, J.: Resilient Data-Centric Storage in Wireless Ad-Hoc Sensor Networks. In: Mobile Data Management (2003)
22. Gibbons, P.B., Matias, Y.: New Sampling-Based Summary Statistics for Improving Approximate Query Answers. In: Proceedings ACM SIGMOD International Conference on Management of Data, Seattle, Washington, pp. 331–342 (June 1998)
23. Gilbert, A., Guha, S., Indyk, P., Kotidis, Y., Muthukrishnan, S., Strauss, M.: Fast, Small-Space Algorithms for Approximate Histogram Maintenance. In: ACM STOC (2002)
24. Gilbert, A., Kotidis, Y., Muthukrishnan, S., Strauss, M.: Optimal and Approximate Computation of Summary Statistics for Range Aggregates. In: ACM PODS, pp. 227–236 (2001)
25. Gilbert, A.C., Kotidis, Y., Muthukrishnan, S., Strauss, M.: One-Pass Wavelet Decompositions of Data Streams. Trans. Knowl. Data Eng. 15(3), 541–554 (2003)
26. He, T., Krishnamurthy, S., Stankovic, J., Abdelzaher, T., Luo, L., Stoleru, R., Yan, T., Gu, L., Hui, J., Krogh, B.: An Energy-Efficient Surveillance System Using Wireless Sensor Networks. In: MobiSys. (2004)
27. Heidermann, J., Silva, F., Intanagonwiwat, C., Govindanand, R., Estrin, D., Ganesan, D.: Building Efficient Wireless Sensor Networks with Low-Level Naming. In: SOSP (2001)
28. Heinzelman, W., Chandrakasan, A., Balakrishnan, H.: Energy-Efficient Communication Protocol for Wireless Microsensor Networks. In: Hawaii Conference on System Sciences (2000)

29. Hellerstein, J.M., Franklin, M.J., Chandrasekaran, S., Descpande, A., Hildrum, K., Madden, S., Raman, V., Shah, M.A.: Adaptive Query Processing: Technology in Evolution. IEEE DE Bulletin 23 (2000)
30. Intanagonwiwat, C., Estrin, D., Govindan, R., Heidermann, J.: Impact of Network Density on Data Aggregation in Wireless Sensor Networks. In: ICDCS (2002)
31. Intanagonwiwat, C., Govindan, R., Estrin, D.: Directed Diffusion: A Scalable and Robust Communication Paradigm for Sensor Networks. In: MOBICOM. (2000)
32. Ioannidis, Y.E., Poosala, V.: Histogram-Based Approximation of Set-Valued Query Answers. In: Proceedings of the 25th VLDB Conference (2000)
33. Kempe, D., Dobra, A., Gehrke, J.: Gossip-Based Computation of Aggregate Information. In: Proceedings of FOCS (2003)
34. Khanna, S., Tan, W.C.: On Computing Functions with Uncertainty. In: Proceedings of ACM PODS Conference (2001)
35. Kotidis, Y.: Snapshot Queries: Towards Data-Centric Sensor Networks. In: Proceedings of ICDE (2005)
36. Kotidis, Y.: Processing Promixity Queries in Sensor Networks. In: Proceedings of DMSN (2006)
37. Lazaridis, I., Mehrotra, S.: Approximate Selection Queries over Imprecise Data. In: ICDE (2004)
38. Lee, J., Kim, D., Chung, C.: Multi-dimensional Selectivity Estimation Using Compressed Histogram Information. In: Proceedings of ACM SIGMOD Conference (1999)
39. Lindsey, S., Raghavendra, C.S.: Pegasis: Power-Efficient Gathering in Sensor Information Systems. In: IEEE Aerospace Conference (2002)
40. Madden, S., Franklin, M.J., Hellerstein, J.M., Hong, W.: Tag: A Tiny Aggregation Service for ad hoc Sensor Networks. In: OSDI Conf. (2002)
41. Madden, S., Franklin, M.J., Hellerstein, J.M., Hong, W.: The Design of an Acquisitional Query processor for Sensor Networks. In: Proceedings of ACM SIGMOD Conference (2003)
42. Mainwaring, A., Polastre, J., Szewczyk, R., Culler, D., Anderson, J.: Wireless Sensor Networks for Habitat Monitoring. In: WSNA 2002, pp. 88–97 (2002)
43. Matias, Y., Vitter, J.S., Wang, M.: Wavelet-Based Histograms for Selectivity Estimation. In: Proceedings of ACM SIGMOD Conference (1998)
44. Motwani, R., Widom, J., Arasu, A., Babcock, B., Babu, S., Datar, M., Manku, G., Olston, C., Rosenstein, J., Varma, R.: Query Processing, Resource Management, and Approximation in a Data Stream Management System. In: Proceedings of CIDR (2003)
45. Olston, C., Jiang, J., Widom, J.: Adaptive Filters for Continuous Queries over Distributed Data Streams. In: Proceedings of ACM SIGMOD Conference (2003)
46. Olston, C., Widom, J.: Offering a Precision-Performance Tradeoff for Aggregation Queries over Replicated Data. In: Proceedings of VLDB (2000)
47. Palpanas, T., Papadopoulos, D., Kalogeraki, V., Gunopulos, D.: Distributed Deviation Detection in Sensor Networks. SIGMOD Rec. 32(4) (2003)
48. Poosala, V., Ioannidis, Y.E.: Selectivity Estimation Without the Attribute Value Independence Assumption. In: Proceedings of the 23th VLDB Conference (1997)
49. Poosala, V., Ioannidis, Y.E., Haas, P.J., Shekita, E.J.: Improved Histograms for Selectivity Estimation of Range Predicates. In: Proceedings of ACM SIGMOD Conference (1996)
50. Priyantha, N.B., Chakraborty, A., Balakrishnan, H.: The Cricket Location-Support System. In: MOBICOM (2000)

51. Ratnasamy, S., Karp, B., Yin, L., Yu, F., Estrin, D., Govindan, R., Shenker, S.: GHT: a Geographic Hash Table for Data-Centric Storage. In: Proceedings of the 1st ACM international workshop on Wireless sensor networks and applications (2002)

52. Savarese, C., Rabaey, J.M., Beutel, J.: Locationing in Distributed Ad-hoc Wireless Sensor Networks. In: ICASSP (2001)

53. Sharaf, A., Beaver, J., Labrinidis, A., Chrysanthis, P.: Balancing Energy Efficiency and Quality of Aggregate Data in Sensor Networks. VLDB Journal (2004)

54. Shnayder, V., Hempstead, M., Chen, B., Allen, G.W., Welsh, M.: Simulating the Power Consumption of Large-Scale Sensor Network Applications. In: Sensys. (2004)

55. Singh, S., Woo, M., Raghavendra, C.S.: Power-Aware Routing in Mobile Ad Hoc Networks. In: ACM/IEEE International Conference on Mobile Computing and Networking (1998)

56. Stollnitz, E.J., DeRose, T.D., Salesin, D.H.: Wavelets for Computer Graphics - Theory and Applications. Morgan Kaufmann Publishers, Inc., San Francisco (1996)

57. Tan, H.O., Korpeoglu, I.: Power Efficient Data Gathering and Aggregation in Wireless Sensor Networks. SIGMOD Record 32(4) (2003)

58. Thaper, N., Guha, S., Indyk, P., Koudas, N.: Dynamic Multidimensional Histograms. In: Proceedings of ACM SIGMOD Conference (2002)

59. Viglas, S.D., Naughton, J.F.: Rate-based Query Optimization for Streaming Information Sources. In: Proceedings of ACM SIGMOD Conference (2002)

60. Vitter, J.S., Wang, M.: Approximate Computation of Multidimensional Aggregates of Sparse Data Using Wavelets. In: Proceedings of ACM SIGMOD Conference (1999)

61. Warneke, B., Last, M., Liebowitz, B., Pister, K.S.J.: Smart Dust: Communicating with a Cubic-Millimeter Computer. IEEE Computer 34(1), 44–51 (2001)

62. Xue, W., Luo, Q., Chen, L., Liu, Y.: Contour Map Matching for Event Detection in Sensor Networks. In: SIGMOD (2006)

63. Yao, Y., Gehrke, J.: The Cougar Approach to In-Network Query Processing in Sensor Networks. SIGMOD Record 31(3), 9–18 (2002)

64. Ye, W., Heidermann, J.: Medium Access Control in Wireless Sensor Networks. Technical report, USC/ISI (2003)

65. Younis, O., Fahmy, S.: HEED: A Hybrid, Energy-Efficient, Distributed Clustering Approach for Ad Hoc Sensor Networks. IEEE Transactions on Mobile Computing 3(4) (2004)

66. Zdonik, S.B., Stonebraker, M., Cherniack, M., Cetintemel, U., Balazinska, M., Balakrishnan, H.: The Aurora and Medusa Projects. IEEE DE Bulletin (2003)

67. Zeinalipour-Yazti, D., Neema, S., Gunopulos, D., Kalogeraki, V., Najjar, W.: Data Acquision in Sensor Networks with Large Memories. In: IEEE International Workshop on Networking Meets Databases, Tokyo, Japan (April 2005)

Load Management and High Availability in the Borealis Distributed Stream Processing Engine

Nesime Tatbul[1], Yanif Ahmad[2], Uğur Çetintemel[2], Jeong-Hyon Hwang[2], Ying Xing[2], and Stan Zdonik[2]

[1] ETH Zürich, Department of Computer Science, Zürich, Switzerland
tatbul@inf.ethz.ch
[2] Brown University, Department of Computer Science, Providence, RI, USA
{yna,ugur,jhhwang,yx,sbz}@cs.brown.edu

Abstract. Borealis is a distributed stream processing engine that has been developed at Brandeis University, Brown University, and MIT. It extends the first generation of data stream processing systems with advanced capabilities such as distributed operation, scalability with time-varying load, high availability against failures, and dynamic data and query modifications. In this paper, we focus on aspects that are related to load management and high availability in Borealis. We describe our algorithms for balanced and resilient load distribution, scalable distributed load shedding, and cooperative and self-configuring high availability. We also present experimental results from our prototype implementation showing the effectiveness of these algorithms.

1 Introduction

In the past several years, data streaming applications have become very common. The broad range of applications include financial data analysis [1], network traffic monitoring [2], sensor-based environmental monitoring [3], GPS-based location tracking [4], RFID-based asset tracking [5], and so forth. These applications typically monitor real-time events and generate high volumes of continuous data at time-varying rates. Distributed stream processing systems have emerged to address the performance and reliability needs of these applications (e.g., [6], [7], [8], [9]).

Borealis is a distributed stream processing engine that has been developed at Brandeis University, Brown University, and MIT. It builds on our earlier research efforts in the area of stream processing - Aurora and Medusa [10]. Aurora provides the core stream processing functionality for Borealis, whereas Medusa enables inter-node communication. Based on the needs of recently emerging stream processing applications, Borealis extends both of these systems in non-trivial and critical ways to provide a number of advanced capabilities. More specifically, Borealis extends the basic Aurora stream processing system with the ability to:

- operate in a distributed fashion,
- dynamically modify various data and query properties without disrupting the system's run-time operation,

S. Nittel, A. Labrinidis, and A. Stefanidis (Eds.): GSN 2006, LNCS 4540, pp. 66–85, 2008.
© Springer-Verlag Berlin Heidelberg 2008

- dynamically optimize processing to scale with changing load and resource availability in a heterogeneous environment, and
- tolerate node and network failures for high availability.

In this paper, we focus on two key aspects of distributed operation in Borealis. Distributing stream processing across multiple machines mainly provides the following benefits:

- **Scalability.** The system can scale up and deal with increasing load or time-varying load spikes with the addition of new computational resources.
- **High availability.** Multiple processing nodes can monitor the system health, and can perform fail-over and recovery in the case of node failures.

In the rest of this paper, we first present a brief overview of the Borealis system. Then we summarize our work on three different aspects of distributed operation in Borealis: load distribution (Section 3), distributed load shedding (Section 4), and high availability (Section 5). Finally, we briefly discuss our plans for future research and conclude.

2 Borealis System Overview

Borealis accepts a collection of continuous queries, represents them as one large network of query operators (also known as a query diagram), and distributes the processing of these queries across multiple server nodes. Sensor networks can also participate in query processing behind a sensor proxy interface which acts as another Borealis node [11].

Queries are defined through a graphical editor, while important run-time statistics such as CPU utilizations of the servers, latencies of the system outputs, and percent data delivery at the outputs are visualized through our performance monitor [12]. Figure 1 provides a snapshot from the editor part of our system GUI.

Each node runs a Borealis server whose major components are shown in Figure 2. The *query processor (QP)* forms the essential piece where local query execution takes place. Most of the core QP functionality is provided by parts inherited from Aurora [13]. *I/O queues* feed input streams into the QP and route tuples between remote Borealis nodes and clients.

The *admin* module is responsible for controlling the local QP, performing tasks such as setting up queries and migrating query diagram fragments. This module also coordinates with the *local optimizer* to provide performance improvements on a running diagram. The local optimizer employs various tactics including, changing local scheduling policies, modifying operator behavior on the fly via special control messages, and locally discarding low-utility tuples via load shedding when the node is overloaded.

The QP also contains the *storage manager*, which is responsible for storage and retrieval of data that flows through the arcs of the local query diagram, including memory buffers and *connection point (CP)* data views. Lastly, the

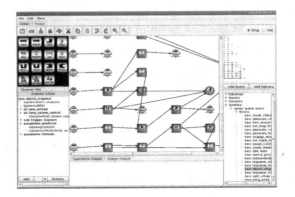

Fig. 1. Borealis query editor

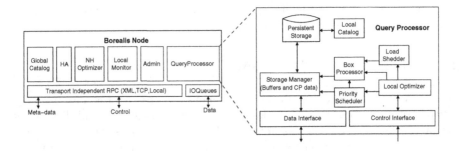

Fig. 2. Borealis system architecture

local catalog stores query diagram description and metadata, and is accessible by all the local components.

Other than the QP, a Borealis node has modules which communicate with their respective peers on other Borealis nodes to take collaborative actions. The *neighborhood optimizer* uses local load information as well as information from other neighborhood optimizers to improve load balance between nodes or to shed load in a coordinated fashion. The *high availability (HA)* modules on different nodes monitor each other and take over processing for one another in case of failures. The *local monitor* collects performance-related statistics as the local system runs to report to local and neighborhood optimizer modules. The *global catalog* provides access to a single logical representation of the complete query diagram.

In addition to the basic node architecture shown in Figure 2, a certain Borealis server can be designated as the coordinator node to perform global system monitoring and to run various global optimization algorithms, such as global load distribution and global load shedding. Thus, Borealis essentially provides a three-tier monitoring and optimization hierarchy (local, neighborhood, and global) that works in a complementary fashion [7].

3   Load Distribution in Borealis

Distributed stream processing engines can process more data at higher speeds by distributing the query load onto multiple servers. The careful mapping of query operators onto available processing nodes is critical in enduring unpredictable load spikes, which otherwise might cause temporary overload and increase in latencies. Thus, the problem involves both coming up with a good initial operator placement as well as dynamically changing this placement as data arrival rates change. Borealis provides two complementary mechanisms to deal with this problem:

- a correlation-based operator distribution algorithm, which exploits the relationship between the load variations of different operators, as well as nodes, in determining and dynamically adjusting the placement of the operators in a balanced way, and

- a resilient operator distribution algorithm, whose primary goal is to provide a static operator placement plan that can withstand the largest possible set of input rate combinations without the need for redistribution.

In this section, we briefly summarize these mechanisms.

3.1   Correlation-Based Operator Distribution

To minimize end-to-end latency in a push-based system such as Borealis, it is important, but not enough, to evenly distribute the average load among the servers. The variation of the load is also a key factor in determining the system performance. For example, consider two operator chains. Each chain consists of two identical operators with cost c and selectivity 1. When the average input rates of the two input streams (r_1 and r_2) are the same, the average loads of all operators are the same. Now consider two operator mapping plans on two nodes. In the first plan, we put each of the two connected operator chains on the same node (Figure 3(a)). In the second plan, we place each operator of a chain on a different node (Figure 3(b)). There is no difference between these two plans from the load balancing point of view. However, suppose that the load bursts of the two input streams happen at different times. For example, assume that $r_1 = r$ when $r_2 = 2r$, or $r_1 = 2r$ when $r_2 = r$. Then these two plans result in very different performance. In the connected plan, there is a clear imbalance between the two nodes in both burst scenarios (Node1's load is $2cr$, when Node2's load is $4cr$, and vice versa); whereas in the cut plan, the load balance is maintained for both of the scenarios (Both nodes have the load of $3cr$ in both cases). Since the two bursts are out of phase, the cut plan which groups operators with low load correlation together, ensures that the load variation on each node is kept small. The simulation result presented in Figure 3(c), which shows that the cut plan can achieve smaller latency with increasing load, also confirms this observation. This simple example clearly shows that, not only the average load, but also the load variation must be considered to achieve a good operator placement that can withstand bursts.

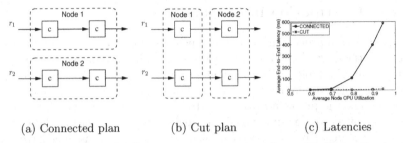

(a) Connected plan   (b) Cut plan   (c) Latencies

Fig. 3. Comparison of different operator mapping plans with fluctuating load

In Borealis, we try to balance the average load among the processing nodes, but we also try to minimize the load variance on each node. For the latter goal, we exploit the correlation of stream rates across the operators. More specifically, we represent operator load as fixed-length time series. The correlation of two time series is measured by the correlation coefficient, which a real number between -1 and 1. Intuitively, when two time series have a positive correlation coefficient, if the value of one time series at a certain index is relatively large in comparison to its mean, then the value of the other time series at the same index also tends to be relatively large. On the other hand, if the correlation coefficient is negative, then when the value of one time series is relatively large, the value of the other tends to be relatively small. Our algorithm is inspired by the observation that if the correlation coefficient of the load time series of two operators is small, then putting these operators together on the same node helps in minimizing the load variance.

The intuition of correlation is also the foundation of the other idea in our algorithm: when making operator allocation decisions, we try to maximize the correlation coefficient between the load statistics of different nodes. This is because moving operators will result in temporary poor performance due to the execution suspension of those operators. However, if the load time series of two nodes have large correlation coefficient, then their load levels are naturally balanced even when the load changes. By maximizing the average load correlation between all node pairs, we can minimize the number of load migrations needed.

As we showed in an earlier paper [14], minimizing the average load variance in fact also helps in maximizing the average load correlation, and vice versa. Therefore, the main goal of our load distribution algorithms is to produce a balanced operator mapping plan where the average operator load variance is minimized or the average node load correlation is maximized. More formally, assume that there are n nodes in the system. Let X_i denote the load time series of node N_i and ρ_{ij} denote the correlation coefficient of X_i and X_j for $1 \leq i, j \leq n$. We want to find an operator mapping plan with the following properties:

- $EX_1 \approx EX_2 \approx ... \approx EX_k$

- $\frac{1}{n} \sum_{i=1}^{n} var X_i$ is minimized, or

- $\sum_{1 \leq i < j \leq n} \rho_{ij}$ is maximized.

Finding the optimal solution to this problem requires comparison of all possible mapping plans and is NP hard. Instead, we developed a number of greedy heuristics which helps us find sub-optimal solutions in polynomial time, and which can experimentally be shown to perform very close to the optimal.

The Borealis coordinator periodically collects load statistics from all nodes, orders nodes by their average load, and pairs them by grouping the ith node with the $(n - i + 1)^{th}$ node in the ordered list. If the load difference between a node pair is above a certain threshold, then operators need to be moved between those nodes to balance their average load in a way that also minimizes their average load variance. Given such a pair, the load movement can be either one-way or two-way:

- In the one-way case, only the more loaded node is allowed to offload half of its excess load to its mate; the purpose is to reduce the load movement overhead. The operators of the more loaded node (say N_1) are ordered based on a score, and the operator with the largest score is moved across to the other node (say N_2) in a greedy fashion until the balance is achieved. The score for an operator o represents the difference between the correlation coefficient between o and the rest of the operators at N_1, and the correlation coefficient between o and the rest of operators at N_2. A larger score makes o a desirable candidate for movement, since this way, the average load variance for the pair can be decreased.

- In the two-way case, all operators on both members of the pair can be moved across freely. Initially, both nodes are treated as empty nodes. At each iteration, we select an operator from the pool of unmapped operators with the largest score and place it at the less loaded node. We continue until all operators have been mapped to one of the nodes. This two-way algorithm can result in a better mapping plan than the one-way algorithm; however, the load movement overhead can be unnecessarily high, especially when the former mapping was relatively good. To address this problem, we add a selective exchange step to our algorithm which would only allow the two-way movement of operators whose score is above certain threshold. By varying this threshold, we can control the tradeoff between the amount of load moved and the quality of the resulting mapping plan.

The above correlation-based load redistribution algorithms can also be modified to handle the case of initial load distribution when all nodes are empty. The algorithm is very similar to the two-way case except that the score formula should be generalized to n nodes rather than considering a single pair. The algorithm in this case is global rather than pair-wise.

In Figure 4, we compare our correlation-based load distribution algorithm against two other load balancing alternatives (randomized load balancing (RAND-GLB) and largest load first load balancing (LLF-GLB)). Figure 4(a) shows that our algorithm maintains low latency with increasing load, and Figure 4(b) confirms that the resulting average load variance is also much smaller (and very close to the optimal) for our algorithm. A detailed description of all of our

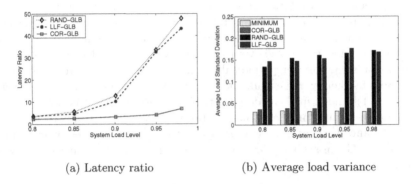

(a) Latency ratio   (b) Average load variance

Fig. 4. Performance comparison for correlation-based global algorithm with others

dynamic load distribution algorithms along with their theoretical and experimental performance evaluation can be found in our earlier work [14].

3.2   Resilient Operator Distribution

Dynamic load distribution techniques described in the previous subsection for balancing load and minimizing latency in the face of unpredictable load variations are more suitable for medium-to-long term load variations, since they persist for relatively longer periods of time and are thus rather easy to capture. Furthermore, the overhead of load redistribution is amortized over time. On the other hand, short-term load fluctuations are both difficult to capture due to their transient nature and too heavy-weight to handle through operator redistribution. To give a concrete example, the base overhead of run-time operator migration in Borealis is measured to be on the order of a few hundred milliseconds (higher for operators with larger state) [15]. Thus, for these kinds of scenarios where operator movement is rather prohibitive, Borealis provides a static resilient operator distribution algorithm.

A resilient operator distribution (ROD) is one that does not become overloaded easily in the face of bursty and fluctuating input rates. This is achieved by optimizing the system to handle as many load points as possible so that it can tolerate those load conditions without the need for operator migration. More specifically, we model the load of each operator as a function of operator costs, selectivities, and system input stream rates. For given input stream rates and a given operator distribution plan, the system is either *feasible* (i.e., none of the nodes are overloaded), or *infeasible* (i.e., at least one node is overloaded). The set of all feasible input rate combinations defines a *feasible set*. Thus, our goal is to find an operator distribution plan that maximizes the size of this feasible set.

Our approach to this problem is based on a linear algebraic model. In this model, we consider a multi-dimensional space of input stream rates, where each processing node is represented by a hyperplane that consists of all input rate points that render this node fully loaded. These node hyperplanes collectively determine the shape and size of the feasible set. Thus, our goal is to find the "ideal" hyperplane which gives us the largest feasible set size. We mathematically showed that this "ideal" feasible set can be achieved if all node hyperplanes are identical (i.e., if the load of each stream is perfectly balanced across all nodes) [15]. However, the ideal feasible set may not always be achievable in practice. Therefore, our main goal is to make the node hyperplanes as close to the ideal hyperplane as possible.

Enumerating all possible operator distribution plans and comparing their feasible set sizes to find an optimal plan is intractable when the number of inputs or the number of operators is large [15]. Therefore, we developed a greedy ROD algorithm which is driven by the following two heuristics:

- **MaxMin Axis Distance (MMAD).** Push the intersection points of the node hyperplanes along each axis, towards those of the ideal hyperplane.

- **MaxMin Plane Distance (MMPD).** Push node hyperplanes directly towards the ideal hyperplane.

Intuitively, MMAD tries to balance the load of each input stream across the nodes in proportion to their CPU capacities, whereas MMPD focuses on the combination of the impact of different input streams on each node to avoid creating bottlenecks at certain nodes. In other words, MMPD tries to balance the load of the nodes in proportion to their CPU capacities for multiple workload points.

The ROD algorithm appropriately combines these heuristics and consists of the following two steps:

- **Operator Ordering.** Sort the operators in descending order of their effect on load.

- **Operator Assignment.** Iteratively assign each operator in the ordered list to a node such that the reduction in the final feasible set size would be minimal. Given an operator o, it is assigned to one of the nodes using a combination of our MMAD and MMPD heuristics. More specifically, at each assignment step, we first separate the nodes into two classes. In Class I, we include those nodes that will not lead to a reduction in the final feasible set size, whereas in Class II, we have the remaining ones. If Class I is not empty, then we choose a node from this class (either randomly or based on another orthogonal criteria [15]), and assign o to this node. Otherwise, o is assigned to the node from Class II which will bear the maximum plane distance. In other words, when we assign operators to Class I nodes, we push the axis intersection points closer to those of the ideal hyperplane as in the MMAD heuristic. On the other hand, when we assign them to Class II nodes, we follow the MMPD heuristic and select the node which has the largest plane distance.

In Figure 5, we show two base results from our performance study on the Borealis prototype running on 10 homogeneous server nodes. We used aggregation-based network traffic monitoring queries. Figure 5(a) compares the feasible set

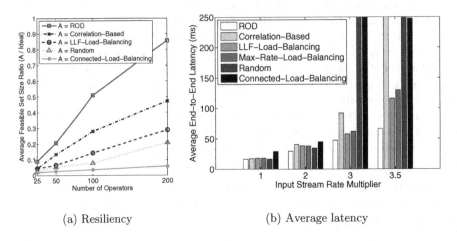

(a) Resiliency   (b) Average latency

Fig. 5. ROD performance

size achieved by different operator distribution algorithms (relative to the ideal). ROD clearly outperforms all of the other alternatives, including our correlation-based load balancing algorithm that was summarized in Section 3.1. As the number of operators increases, ROD approaches to the ideal case and most of the other algorithms also improve because there is a greater chance that the load of a given input stream will be spread across multiple nodes. Figure 5(b) compares average end-to-end latency achieved as a result of applying various load distribution algorithms as the CPU utilization is increased from 26% to 79% (corresponding to input rates multipliers of 1 and 3.5, respectively). Since ROD produces the largest feasible set size and since it balances the node loads considering multiple input rate combinations, it performs and scales better than the other alternatives. Our results demonstrate that, for a representative workload and data set, ROD (i) sustains longer and is more resilient than the alternatives, and (ii) despite its high resiliency, it does not sacrifice latency performance.

We have also extended our ROD algorithm to handle nonlinear load models, to exploit additional workload information, and to consider communication costs. Details of ROD and its extensions together with their detailed performance results can be found in our earlier work [15].

4   Distributed Load Shedding in Borealis

Data streams can arrive in bursts and provisioning the system for the worst-case load (which can be orders of magnitude higher than the average load) is in general not economically sensible. On the other hand, bursts in data rates may create overload on servers which slows down processing and causes delayed outputs. This is unacceptable in terms of quality of service of real-time streaming applications, where low-latency is a major requirement. Borealis provides load shedding techniques to make sure that all servers always operate below their processing capacity limits. This is achieved by inserting load reducing drop operators at selected arcs of the query network. Dropped tuples result in approximate answers. Therefore, the main goal in our load shedding algorithms is to minimize the degradation in answer quality [1].

In a distributed stream processing system, each node acts like a workload generator for its downstream nodes. Therefore, resource management decisions at any node will affect the characteristics of the workload received by its children. Because of this load dependency between nodes, a given node must figure out the effect of its load shedding actions on the load levels of its descendant nodes. Load shedding actions at all nodes along the chain will collectively determine the quality degradation at the outputs. This makes the problem more challenging than its centralized counterpart [16].

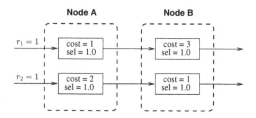

Fig. 6. Motivating example

To illustrate, consider the simple query network in Figure 6, with two queries that are distributed onto two processing nodes A and B. Each small box represents a subquery with a certain cost and selectivity. Cost reflects the CPU time that it takes for one tuple to be processed by the subquery, and selectivity represents the ratio of the number of output tuples to the number of input tuples. Both inputs arrive at the rate of 1 tuple per second. Potentially each node can reduce load at its inputs by dropping tuples to avoid overload. Let's consider node A. Table 1 shows various ways that A can reduce its input rates and the consequences of this in terms of the load at both A and B, as well as the throughput observed at the query outputs (Note that we are assuming a fair scheduler that allocates CPU cycles among the subqueries in a round-robin fashion). In all of these plans, A can reduce its load to the capacity limit. However, the effect of each plan on B can be very different. In plan 1, B stays at the same overload level. In plan 2, B's load increases to more than twice its original load. In plan 3, B's overload problem is also resolved, but throughput is low. There is a better plan which removes overload from both A and B, while delivering the highest total throughput (plan 4). However, node A can only implement this plan if it knows about the load constraints of B. From A's point of view, the best local

---

[1] In this work, we focus on total query throughput as the quality metric to maximize.

Table 1. Alternate load shedding plans for node A of Figure 6

Plan	Reduced rates at A	A.load	A.throughput	B.load	B.throughput	Result
0	1, 1	3	1/3, 1/3	4/3	1/4, 1/4	originally, both nodes are overloaded
1	1/3, 1/3	1	1/3, 1/3	4/3	1/4, 1/4	B is still overloaded
2	1, 0	1	1, 0	3	1/3, 0	optimal plan for A, but increases B.load
3	0, 1/2	1	0, 1/2	1/2	0, 1/2	both nodes ok, but not optimal
4	1/5, 2/5	1	1/5, 2/5	1	1/5, 2/5	optimal

plan is plan 2. This simple example clearly shows that nodes must coordinate in their load shedding decisions to be able to achieve high-quality query results.

We model the distributed load shedding problem as a linear optimization problem. In our formulation, each server node is represented with a linear load constraint, written in terms of operator costs, selectivities, and input rates. The objective function to maximize is the total output rate at the query end-points, written in terms of operator selectivities and input rates. The drop selectivities (i.e., the fraction of tuples to be kept at the designated drop arcs) appear as the variables in both of these formulas. The goal is to solve the linear program to assign the optimal values to these variables that would satisfy the load constraints on all servers while maximizing the total throughput objective [17].

Our solution to the distributed load shedding problem consists of four steps: (i) advanced planning, (ii) load monitoring, (iii) plan selection, and (iv) plan implementation. In the first step, we precompute a series of load shedding plans for various input rate combinations, each corresponding to an overload condition. The idea is to prepare the system against any potential overload scenario by doing most of the computational work in advance. Next we start periodically measuring the system load. If an overload is detected in one or more of the servers, we select a plan from the previously computed ones and modify the query network according to this plan.

We architect our solution in two alternative ways:

- **Centralized Approach.** In the centralized solution, all load shedding steps are performed at one central server (designated as the "coordinator node") except the plan implementation step. The coordinator contacts all the other servers in order to collect information on their query network topology and run-time statistics (e.g., operator costs and selectivities). Based on the collected global metadata, the coordinator generates a series of load shedding plans for other servers to apply under certain overload conditions. Here, we use the GNU Linear Programming Toolkit (GLPK) [2] to generate the plans. These plans are then uploaded onto the associated servers together with their plan-id's. Then the coordinator starts monitoring the input load. If an overload situation is detected, the coordinator selects the best plan to apply and sends the corresponding plan-id to the other servers in order to trigger the distributed implementation of the selected plan.

---

[2] http://www.gnu.org/software/glpk/glpk.html

- **Distributed Approach.** In the distributed solution, all four load shedding steps are performed at all of the participating nodes in a cooperative fashion. The collective actions of all the servers result in a globally effective load shedding plan. The neighboring servers coordinate through metadata aggregation and propagation. As a result of this communication, each node identifies what makes a feasible input load for itself and its server subtree, and represents this information in a table that we call the *Feasible Input Table (FIT)*. FIT is then propagated to the upstream parent. The parent aggregates the FITs from all of its children, eliminating the table entries that are infeasible for itself. Finally, the parent propagates the resulting FIT to its own parents. This propagation continues until the input nodes receive the FIT for all their downstream nodes. Then using its FIT, a node can shed load for itself and for its descendant nodes.

In the rest of this section, we describe how we perform the advance planning step for the above alternative approaches.

4.1   Solver-Based Advance Planning

Our goal in the advance planning step of the centralized, solver-based approach is to produce load shedding plans for a set of infeasible input rate combinations, which will make them feasible for all the servers in the system. The number of such combinations to consider could be potentially very large, and it would be too costly to call the LP solver for each such combination. Instead we use the following, more efficient strategy: We consider a multi-dimensional space of input rates. We systematically search this space to pick a subset of the possible points for which we will call the solver. For the rest of the points, we approximately reuse the solver-generated plans. To be more specific, we assume that an error threshold in quality, ϵ is defined. Given any infeasible point s that lies between two other infeasible points r and q (i.e., $r < s < q$) for which we have already computed the optimal load shedding plan using the solver, if $(q.quality - r.quality) \leq \epsilon * q.quality$, then s can use the plan for r with a minor modification. For example, consider the two-dimensional example in Figure 7(a), each dimension representing an input stream. Assume that $s = (60, 75)$ lies within ϵ-distance from $r = (50, 50)$ and $q = (100, 100)$. Then s can use the plan at r, with the additional modification that input1 and input2 must be reduced by an additional factor of $\frac{50}{60}$ and $\frac{50}{75}$, respectively. Based on this idea, we take the input rate space and divide it in a region quadtree-like fashion. For each region, the solver is called for the top-most and the bottom-most points. We stop dividing a region either when all points in that region turn out to be within the ϵ bound, or the top-most point is in fact a feasible point. The result of this process is a collection of input rate subspaces with a load shedding plan assigned to each subspace. These subspaces can be very conveniently placed into a quadtree-based index during the space division process described above. For example, Figure 7(b) shows the index that corresponds to the space division of Figure 7(a). At run time, we will use this index to locate the region into which an

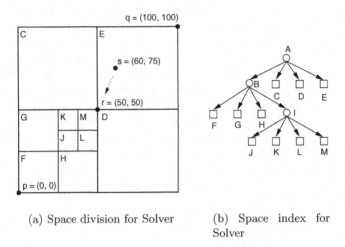

(a) Space division for Solver   (b) Space index for Solver

Fig. 7. Quadtree-based space division and index for Solver

observed infeasible rate point falls and will use the corresponding load shedding plan.

4.2   FIT-Based Advance Planning

Our goal in the advance planning step of the distributed [3], FIT-based approach is to represent a set of feasible input rate combinations (for each server and its subtree) with a table. To briefly summarize, given a node with m inputs, the FIT for this node is a table with $m + 2$ columns. The first m columns represent the rates for the m inputs; the $(m+1)$th column represents the *complementary local load shedding plan* that must be used together with that input entry (this plan may be needed to handle query sharing [17]); and the last column represents the resulting output quality score. Again, for efficiency, we do not want to consider all possible rate combinations. Instead, we use the ϵ threshold as follows: From each input dimension, we pick FIT points that are at most some distance apart, we call this distance "spread" for that dimension. If input dimension i has a maximum feasible rate m_i, then the spread for that dimension can be computed as $\epsilon * m_i$. Next, we must map potential infeasible points to our feasible points in FIT. This means that, if we observe a certain infeasible point q, then we will reduce it to a feasible point p using proper drop values. In general, an infeasible point q that is greater than a feasible point p on all dimensions can be mapped to p. However, we would like to use the best mapping. Our algorithm divides the input rate space accordingly to make sure that this assignment is done to guarantee the highest quality for all infeasible points. The resulting set of subspaces are again placed in a quadtree-based data structure to facilitate

---

[3] Although FIT is a distributed algorithm by design, its centralized implementation is also available [17].

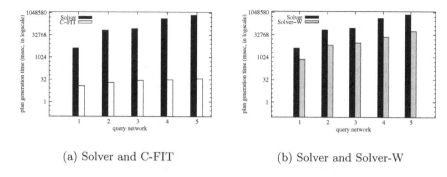

(a) Solver and C-FIT   (b) Solver and Solver-W

Fig. 8. Plan generation performance for different query networks

search at run-time. Further details about how FIT points are generated and how this table is propagated between neighboring nodes are described in our earlier work [17], [18].

In Figure 8, we show a basic experimental result that compares the load shedding plan generation time for our alternative approaches on five different query networks. These networks differ in the way they apportion the query load across different query paths. In Figure 8(a), we are showing that centralized implementation of FIT outperforms Solver and it is also less sensitive to query load imbalance. The performance difference is mainly due to the time Solver spends in searching the space of infeasible points, while FIT only deals with the feasible points. In Figure 8(b), we compare Solver with its variation Solver-W. Solver-W essentially takes workload knowledge into account and tries to meet a given expected error threshold for the average case. In other words, some infeasible points are known to be less likely than others. Errors of such points contribute less to the average error. Therefore, the algorithm spends less time exploring the corresponding subspaces. The result is some improvement in plan generation performance. Further experimental results can be found in our previous work [17].

5   High-Availability in Borealis

In a distributed stream processing system, servers may fail and this can significantly disrupt or even halt overall stream processing. In case of failures, large amount of transient information may be lost and the servers downstream from a failed one may stop making any progress. Therefore, a distributed stream processing system must provide high-availability (HA) mechanisms that allows processing to continue in spite of server failures. These mechanisms must take correctness (e.g., data loss, duplicates) and performance (e.g., latency introduced during regular processing and during recovery time) requirements of the applications into account.

5.1   Basic HA Models

In Borealis, we define three types of recovery guarantees to address different HA requirements:

- **Precise Recovery.** Post-failure output is exactly the same as the output without failure. Many financial services applications have such strict correctness requirements.
- **Rollback Recovery.** The output produced after a failure is "equivalent" to that of *an execution* without failure, but not necessarily to the output of *the failed execution*. The output may also contain duplicate tuples. Thus, information loss is avoided, but the output can still be imprecise. Event detection applications such as fire alarms, theft prevention are examples.
- **Gap Recovery.** This is the weakest form of recovery where data loss is acceptable for better performance. Sensor-based environment monitoring where recent data is more important is an example.

Each primary server has an associated backup server. A backup server runs its own stream processing engine and has the same query network fragment as its primary, but its state is not necessarily the same as that of the primary. If a primary server fails, its backup server immediately detects the failure and takes over the operation of the failed server.

Borealis provides four recovery approaches that can provide one or more of the above recovery guarantees. These approaches mainly differ in how primary and backup servers prepare for failures. Each approach uses a different combination of redundant processing, checkpointing, and remote logging. As a result, they offer different tradeoffs between run-time overhead and recovery performance.

- **Amnesia.** This approach does not involve any preparation for failures. As soon as the backup server detects that the primary has failed, it restarts the failed query network from an empty state.
- **Passive Standby.** Each primary server periodically checkpoints (i.e., reflects its state updates) to its backup server. The backup server takes over from the latest checkpoint when the primary fails.
- **Active Standby.** The backup server processes all tuples in parallel with its primary. The output tuples of the backup server are not sent downstream; instead they are logged at the output queues. If the primary fails, the backup takes over by sending the logged tuples to all downstream neighbors and then continuing its processing.
- **Upstream Backup.** Upstream servers preserve tuples in their output queues while their downstream neighbors are still processing them. If a server fails, an empty backup server rebuilds the latest state of the failed primary from the logs kept at the upstream server.

The amnesia approach can only provide gap recovery guarantee, while the other approaches provide rollback recovery in their simplest forms and can be extended to provide precise recovery. In principle, the guarantee of precise recovery requires a higher run-time cost than other weaker recovery guarantees. Furthermore, the query operators may also affect recovery semantics and associated cost requirements. Some Borealis operators are deterministic (i.e., they produce the same output stream every time they start from the same initial state and receive the same input tuples), while others are arbitrary due to dependence on time or arrival order. Thus, deterministic ones are less costly to provide better guarantees [19].

An in-depth algorithmic analysis of all of the above basic HA alternatives together with results from our experimental study showed that each HA approach poses a clear tradeoff between recovery time and processing overhead [19]. In fact, each approach covers a complementary portion of the solution space. To summarize:

- Active standby has high run-time overhead, but provides very fast recovery.

- Passive standby performs worse than active standby both in terms of recovery time and run-time overhead. However, it is the only approach that easily provides precise recovery for arbitrary query networks. Additionally, it can flexibly trade off between run-time overhead and recovery speed by adjusting the checkpoint interval.

- Upstream backup provides precise recovery for most query networks with the lowest run-time overhead, but at the cost of a longer recovery, depending on the amount of logged data to process during recovery.

5.2   Cooperative and Self-configuring HA for Server Clusters

A server cluster is a popular form of shared-nothing computing architecture where commodity servers are connected by fast local area networks. Borealis may distribute its processing load onto such a cluster for better scalability. For such environments, we designed and implemented a self-configuring HA approach that enables fast recovery as well as minimal slow-down for regular processing. Unlike our basic HA mechanism described in Section 5.1 where each server is assigned one other backup server, in this case, each server is backed up by multiple servers in a cooperative fashion. Each of these backup servers are in charge of a disjoint query network fragment (called an "HA unit") of the primary server. Thus, they can take over the failed execution in parallel, which speeds up the recovery time of rebuilding the latest state of the failed server. Furthermore, HA tasks are performed when servers are idle, which reduces the interference with regular stream processing.

In this work, we focused on checkpoint-based passive standby as the recovery approach. This choice is mainly due to the fact that checkpointing works for a larger set of workload and usage scenarios than the other alternatives. Below we briefly summarize the important features of this approach; more technical details can be found in our previous work [20].

- **The HA Mechanism.** Query network on each server is partitioned into HA units. Each such unit is assigned to a different backup server. The preparation for failures involves two HA tasks, namely *capture* and *paste*, to be performed during idle periods. Capture is performed by the primary server, while paste is performed by the backup server. In capture, the primary selects one of its HA units, prepares a checkpoint message for it that includes all the state changes since the last checkpoint, and sends this message to the associated backup server. In paste, the backup selects one of the checkpoint messages that it has received from a primary, copies the message to the corresponding backup image, and notifies the sender primary that the checkpointing request has been completed. Each server is periodically pinged by another designated server. If a failure is detected, then this is broadcast to other servers in the cluster. Each of these notified servers immediately pastes any checkpoint messages from the failed server to the corresponding backup images. Then the execution of these backup images start while the necessary input and output streams are redirected so that stream processing can continue at the backup servers. This HA mechanism provides precise recovery because each backup image can obtain the tuples that the primary has processed since the last checkpoint. This is achieved by keeping output queues at the output of each HA unit to retain those tuples that the downstream backups are currently missing. These output queues are pruned when the downstream server processes them and checkpoints the effect onto the backup server.

- **Checkpoint Scheduling.** A server can be a primary for some HA units and can be a backup for others. Therefore, when it is idle, it can perform either a capture task, or a paste task. The recovery time can be significantly reduced by a careful scheduling of these HA tasks. We developed an algorithm called the "Min-Max Checkpoint Scheduling Algorithm". The idea is to schedule the HA task that would minimize the maximum recovery load among the ones that are in the task queue. This algorithm first finds the best capture task, i.e., the capture of the HA unit with high processing load and low checkpointing cost. Similarly, it finds the best paste task that would help the HA unit with the largest recovery load. Finally, it performs the best task found.

- **Dynamic Backup Assignment.** Assignment of HA units to backup servers can also affect the recovery time. For example, a server which is assigned too many HA units for backup may become a bottleneck. Furthermore, an existing backup assignment may need to be changed with varying system conditions (e.g., changing input rates). Therefore, our approach periodically runs a "Backup Reassignment Algorithm". This algorithm detects *the worst point of failure* (i.e., the server whose failure would cause the longest recovery), and balances its backup load with another server whose failure would cause the shortest recovery.

- **Delta Checkpointing.** HA tasks can be performed more efficiently using operator-specific delta-checkpointing techniques. This is important for stateful operators such as aggregate and join. We use dirty bits for aggregate groups and windows to mark whether they were created after the last checkpoint or not. Dirty windows are fully captured/pasted while others are

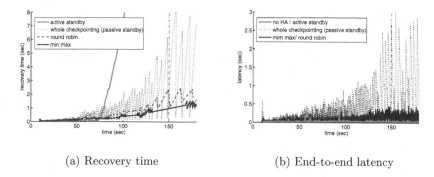

(a) Recovery time   (b) End-to-end latency

Fig. 9. Performance of min-max checkpoint scheduling

partially captured/pasted. For join, only the tuples that entered the window after the last checkpoint are captured.

We performed various experiments on the Borealis prototype in order to evaluate the performance of the above techniques [20]. Figure 9 is a basic result that shows how our min-max checkpoint scheduling algorithm effectively reduces the recovery time while being minimally intrusive to regular query processing. In this experiment, 16 aggregates were deployed on each of 5 identical servers, and input stream rates were increased at time point 150 seconds, when each server became around 90% utilized for query processing. Figure 9(a) shows that min-max algorithm provides the fastest recovery even after the system load is increased. Figure 9(b) shows how HA tasks affect query processing performance. Our finer-grained checkpoint technique disrupts processing much less than the standard whole checkpointing approach.

6   Conclusions and Future Work

This paper provides an overview of the Borealis system and three of its features that are key for its scalable and reliable distributed operation. With our resilient operator distribution algorithm, Borealis can withstand high degrees of load without the need for any operator migration. Beyond that, our correlation-based operator distribution algorithm can dynamically balance server loads by taking the relationship between load variations of operators and nodes into account. Our work on distributed load shedding has focused on the load dependency between different servers, and has proposed two alternative solution architectures for removing CPU overload, where scalable coordination between neighboring servers can be achieved in a centralized or a distributed way. Finally, our work on high availability has explored various recovery guarantees and models that may be demanded by different applications, and has shown the existing tradeoffs between performance and correctness. This work has further explored efficient checkpoint-based recovery techniques for server clusters based on cooperation among multiple servers and automatic self-configuration with changing load. All of these algorithms have been implemented and experimentally evaluated on our system prototype. The latest Borealis prototype code can be downloaded from http://www.cs.brown.edu/research/borealis/.

We are currently working on a replication-based stream processing scheme that will provide Borealis with faster and more reliable operation over wide-area networks [21]. Other future work items include support for richer data types (such as video streams) in the form of multi-dimensional arrays, and seamless integration of stream processing with large-scale data collection and dissemination.

Acknowledgements. We thank all members of the Borealis project for their support. This research has been sponsored by the NSF under the grants IIS-0086057 and IIS-0325838.

References

1. Whitney, A.T., Shasha, D.: Lots o' Ticks: Real-Time High Performance Time Series Queries on Billions of Trades and Quotes (Demo). In: ACM SIGMOD Conference, Santa Barbara, CA (2001)
2. Babu, S., Subramanian, L., Widom, J.: A Data Stream Management System for Network Traffic Management. In: ACM Workshop on Network-Related Data Management (NRDM), Santa Barbara, CA (2001)
3. Stefanidis, A., Nittel, S. (eds.): Geosensor Networks. CRC Press, Boca Raton (2004)
4. Leonhardt, U., Magee, J.: Multi-sensor Location Tracking. In: International Conference on Mobile Computing and Networking (MobiCom), Dallas, TX (1998)
5. Franklin, M.J., Jeffery, S.R., Krishnamurthy, S., Reiss, F., Rizvi, S., Wu, E., Cooper, O., Edakkunni, A., Hong, W.: Design Considerations for High Fan-In Systems: The HiFi Approach. In: CIDR Conference, Asilomar, CA (2005)
6. Shah, M.A., Hellerstein, J.M., Brewer, E.: Highly-Available, Fault-Tolerant, Parallel Dataflows. In: ACM SIGMOD Conference, Paris, France (2004)
7. Abadi, D., Ahmad, Y., Balazinska, M., Çetintemel, U., Cherniack, M., Hwang, J., Lindner, W., Maskey, A., Rasin, A., Ryvkina, E., Tatbul, N., Xing, Y., Zdonik, S.: The Design of the Borealis Stream Processing Engine. In: CIDR Conference, Asilomar, CA (2005)
8. Pietzuch, P., Ledlie, J., Shneidman, J., Roussopoulos, M., Welsh, M., Seltzer, M.: Network-Aware Operator Placement for Stream-Processing Systems. In: IEEE ICDE Conference, Atlanta, GA (2006)
9. Amini, L., Jain, N., Sehgal, A., Silber, J., Verscheure, O.: Adaptive Control of Extreme-scale Stream Processing Systems. In: IEEE ICDCS Conference, Lisboa, Portugal (2006)
10. Zdonik, S., Stonebraker, M., Cherniack, M., Çetintemel, U., Balazinska, M., Balakrishnan, H.: The Aurora and Medusa Projects. IEEE Data Engineering Bulletin (Special Issue on Data Stream Processing) 26 (2003)
11. Abadi, D., Lindner, W., Madden, S., Schuler, J.: An Integration Framework for Sensor Networks and Data Stream Management Systems (Demo). In: VLDB Conference, Toronto, Canada (2004)

12. Ahmad, Y., Berg, B., Çetintemel, U., Humphrey, M., Hwang, J., Jhingran, A., Maskey, A., Papaemmanouil, O., Rasin, A., Tatbul, N., Xing, W., Xing, Y., Zdonik, S.: Distributed Operation in the Borealis Stream Processing Engine (Demo). In: ACM SIGMOD Conference, Baltimore, MD (2005)
13. Abadi, D., Carney, D., Çetintemel, U., Cherniack, M., Convey, C., Lee, S., Stonebraker, M., Tatbul, N., Zdonik, S.: Aurora: A New Model and Architecture for Data Stream Management. VLDB Journal 12 (2003)
14. Xing, Y., Zdonik, S., Hwang, J.H.: Dynamic Load Distribution in the Borealis Stream Processor. In: IEEE ICDE Conference, Tokyo, Japan (2005)
15. Xing, Y., Hwang, J.H., Çetintemel, U., Zdonik, S.: Providing Resiliency to Load Variations in Distributed Stream Processing. In: VLDB Conference, Seoul, Korea (2006)
16. Tatbul, N., Çetintemel, U., Zdonik, S., Cherniack, M., Stonebraker, M.: Load Shedding in a Data Stream Manager. In: VLDB Conference, Berlin, Germany (2003)
17. Tatbul, N., Çetintemel, U., Zdonik, S.: Staying FIT: Scalable Load Shedding Techniques for Distributed Stream Processing. Technical Report CS-06-13, Brown University, Computer Science (2006)
18. Tatbul, N., Zdonik, S.: Dealing with Overload in Distributed Stream Processing Systems. In: IEEE International Workshop on Networking Meets Databases (NetDB), Atlanta, GA (2006)
19. Hwang, J.H., Balazinska, M., Rasin, A., Çetintemel, U., Stonebraker, M., Zdonik, S.: High-Availability Algorithms for Distributed Stream Processing. In: IEEE ICDE Conference, Tokyo, Japan (2005)
20. Hwang, J.H., Xing, Y., Çetintemel, U., Zdonik, S.: A Cooperative, Self-Configuring High-Availability Solution for Stream Processing. In: IEEE ICDE Conference, Istanbul, Turkey (2007)
21. Hwang, J.H., Çetintemel, U., Zdonik, S.: Fast and Reliable Stream Processing over Wide Area Networks. In: IEEE International Workshop on Scalable Stream Processing Systems (SSPS), Istanbul, Turkey (2007)

Knowledge Aquisition and Data Storage in Mobile GeoSensor Networks

Peggy Agouris[1], Dimitrios Gunopulos[2], Vana Kalogeraki[2], and Anthony Stefanidis[1]

[1] Department of Geography and Geoinformation Sciences
George Mason University
Fairfax, VA 22030
{pagouris,astefani}@gmu.edu
[2] Department of Computer Science and Engineering
University of California, Riverside
Riverside, CA 92521
{dg,vana}@cs.ucr.edu

Abstract. In this paper we address the issue of mobility in geosensor networks, inspired by the computational challenges imposed by modern surveillance applications. More specifically we consider networks of optical sensors (video and still cameras), and present a spatiotemporal framework for the management of information captured in them. In this context, mobility is addressed at two levels, considering mobile objects in the area monitored by a network, and mobile sensors observing such objects. Our interest lies on the data acquisition and storage problems that arise in this setting. We identify certain key issues behind the development of a general framework for knowledge acquisition and data storage in geosensor networks, namely: spatiotemporal object modeling; similarity metrics to compare spatiotemporal objects; storing and indexing spatiotemporal objects in a geosensor network; and network management using spatiotemporal techniques. We present some emerging approaches that address these key issues and thus outline a general framework for information and sensor management in mobile sensor networks.

Keywords: Mobility, surveillance, modeling, spatiotemporal similarity, indexing.

1 Introduction

Geosensor networks are emerging as a novel paradigm for geospatial information collection and management. A geosensor network can be loosely defined as a sensor network that monitors phenomena in geographic space, and in which the geospatial content of the information collected, aggregated, analyzed, and monitored is of prime importance [Nittel & Stefanidis, 2004]. For example, cameras and GPS sensors onboard static or mobile platforms have the ability to provide continuous streams of geospatially-rich information. The geographic space covered by a network, and analyzed through its measurements, may range in scale from the confined environment of a room [Chen et al., 2002] or a workplace environment [Conner et al., 2005] to the highly complex dynamics of an ecosystem region [Ailamaki et al., 2003; Juang et al., 2002; Mainwaring et al., 2002].

S. Nittel, A. Labrinidis, and A. Stefanidis (Eds.): GSN 2006, LNCS 4540, pp. 86–108, 2008.

In this paper we address the issue of mobility in geosensor networks, inspired by the computational challenges imposed by modern surveillance applications. More specifically we consider networks of optical sensors (video and still cameras), and present a spatiotemporal framework for the management of information captured in them. In this context, mobility is addressed at two levels, considering *mobile objects* in the area monitored by a network, and *mobile sensors* observing such objects. We do not focus on individual sensor data processing techniques; we assume instead that different sensors provide their readings to the system in the form of a stream (of values or events) and we focus instead on the storage and management of such data in a network of many sensors. Our interest lies on the data acquisition and storage problems that arise in this setting. We identify certain key issues behind the development of a general framework for knowledge acquisition and data storage in geosensor networks with mobility in both the sensor and the tracked object level. They are: spatiotemporal object modeling; similarity metrics to compare spatiotemporal objects; storing and indexing spatiotemporal objects in a geosensor network; and network management using spatiotemporal techniques. In this position paper we present some emerging approaches that address these key issues and thus outline a general framework for information and sensor management in mobile sensor networks.

Even though our motivation stems from video tracking applications, the problems we address are in principle applicable to a wide array of spatiotemporal datasets (e.g. to process information collected using GPS-enabled cell phones, or even RFID tags). The paper is organized as follows. In Chapter 2 we present an overview of relevant literature, followed in Chapter 3 by our approach to model spatiotemporal information in helixes. Chapter 4 presents spatiotemporal similarity assessment techniques to support the comparison of spatiotemporal activities, followed by appropriate indexing techniques as they are presented in Chapter 5. Chapter 6 discusses the issue of distributed storage of spatiotemporal information and in-network analysis for real-time tracking. We conclude with closing remarks in Chapter 7.

2 Related Work

The use of video sensors, tracking devices, and sensor networks is revolutionizing geospatial information collection and analysis, by supporting the capturing of spatiotemporal movement and complex activities. Furthermore, the emergence of sensor networks has introduced some interesting new challenges and approaches to the management of rapidly evolving distributed information. Object tracking in geospatial applications has been greatly assisted over the past few years mainly by advancements in two complementary fields:

- technological advancements in global positioning system (GPS) and relevant tracking technology (e.g. Radio Frequency Identification –RFID- systems) have resulted in the development of economical and easily deployable devices to directly collect positional information over time (coordinates of points in a georeferenced coordinate system, see e.g. [Nascimento et al., 2003]), while
- theoretical advancements in image analysis have resulted in efficient algorithms to track objects as they are captured in motion imagery datasets (i.e. video feeds or sequences of static images).

Early efforts addressed traffic monitoring using stationary cameras [Beymer et al., 1997]. The extension of computer vision solutions multi-view surveillance using motion imagery in large and complex environments was a natural progression [Collins et al., 2001]. Representative efforts were the ones performed in the context of DARPA's VIVID program to model basic patterns of activity [Stauffer & Grimson, 2000] and develop hierarchical video event ontologies [Bolles & Nevatia, 2004].

Addressing *multi-camera systems*, in the context of computer vision applications, the main challenge is the linking of information across feeds. Towards this goal, we have efforts that proceed using a single [Altenis & Jensen, 2002] or limited number of attributes [Eltoukhy & Salama, 2001], and efforts concentrating on overlapping views [Javed et al., 2000]. Of particular interest is the development of correspondence models for overlapping camera networks [Stauffer & Tieu, 2003] or non-overlapping networks in which the movement is constrained to nearly linear patterns [Javed et al., 2000]. An interesting approach based on Hidden Markov descriptions of positions and traffic flow for object linking across various static sensors was presented in [Jaynes, 2004]. [Chang et al., 2000] use epipolar geometry to constrain correspondences across camera views, and use a Bayes net to combine features (e.g., color, height) for correspondence assessment, similar to [Dockstader & Tekalp 2001].

Addressing the *labeling and comparison of trajectories,* we have the development of distance metrics to compare trajectories of equal duration [Makris & Ellis, 2002; Jaynes et al., 2002], the introduction of spatial and temporal shifts in this comparison [Needham & Boyle, 2003], and the use of distance metrics that are based on model-based trajectory representations to compare trajectories of varying resolutions [Porikli, 2004]. Activity detection in the presence of noise using optical and infrared cameras, and the classification of activities in rather constrained environments has also been addressed in specific applications, and typically under highly constrained conditions (e.g. monitoring vessels as they enter ports [Rhodes et al., 2005]), or work on object tracking through enter/exit zones.

Addressing *target tracking in sensor networks*, centralized approaches for predicting a moving object's location have been proposed in the literature. In [Aslam et al, 2003] a binary model for tracking a moving object is used, in which sensors employ a single bit of information (that indicates whether the object is approaching or moving away from a sensor) and broadcast this bit to a base station that can accurately estimate the object's trajectory. The need to find ways to balance energy consumption and accurate object prediction has been recently addressed. Approaches have been proposed for clustering the sensors into groups to minimize the network's energy consumption [Bandyopadhyay & Coyle, 2003] and to efficiently disseminate data taking into consideration account sink mobility [Ye et al, 2002]. In these, the target's location can be predicted based on known previous locations [Yand & Sikdar, 2003], or through the collaboration of multiple nodes surrounding the target to increase accuracy and fault tolerance [Cerpa et al, 2001]. To save network resources, [Brooks et al, 2003] use a collaborative signal processing (CSP) approach based on location-aware data routing that limits the scope of CSP to the relevant subset of nodes only. In [Zhang & Gao, 2004] the *Dynamic Convoy Tree-based Collaboration* (DCTC) framework builds a tree structure called convoy tree, with sensor nodes around the moving target. As the target moves, the tree dynamically evolves by adding new nodes and pruning others. In this case, many nodes in the convoy tree may become far away from the root of the tree. As a result, a new root is elected to replace the old root, and the tree is reconfigured accordingly.

A new class of spatiotemporal multicast protocols have been recently developed [Huang et al, 2003(a); Huang et al, 2003(b)] that take into account spatio-temporal constraints to provide reliable and just-in-time delivery. In [Liu et al, 2003], geographically local collaborative groups are formed and the tradeoff between performance and scalability for target localization is discussed. In [Abdelzaher et al, 2004], object-oriented programming mechanisms have been proposed for implementing tracking applications. The focus is on hiding from the application developer the issues of object communication, object mobility and the maintenance of the tracking objects and their states.

Work relevant to sensor information management, querying, and integration has been performed in *data streams*, with notable projects like Stanford's *STREAM* [Motwani 2003], Berkeley's *Telegraph* [Madden et al. 2002], MIT/Brown's *Aurora* [Carney et al. 2002], and Cornell's *Cougar* [Faradjian et al. 2002]. The *STREAM* project made significant contributions to formalizing general data streams. *Telegraph*, *Aurora*, and *Cougar* have focused on disseminating queries and collecting query results over sensor networks which consists of sets of energy-constrained, small-form sensor platforms. This work also includes approximate queries on data streams using statistical sampling techniques [Gehrke et al. 2001]. UCLA's Smart Kindergarten Project [Srivastava 2001] uses sensors like video cameras, microphones and others; data collected in a playroom setting and combined with data from embedded sensors in 'smart toys'. Here, data streams are annotated with context information to provide for improved analysis capabilities of the video data.

3 Modeling Spatiotemporal Trajectories in ST Helixes

The spatiotemporal evolution of an object comprises two types of activities: *movement* and *deformation*. As the object *moves,* it changes its location with respect to an external reference frame. This information can be represented as a trajectory describing the movement of the object's center of mass. The second type of spatiotemporal change is deformation. It describes the variations of the object's shape with respect to an internal reference frame. In order to describe both of these types of change we have developed the model of the spatiotemporal helix [Agouris & Stefanidis, 2003; Stefanidis et al., 2003]. A visualization of a helix is shown in Fig. 1, using the 3-dimensional *(x,y,t)* spatiotemporal domain of a scene, comprising two *(x, y)* spatial dimensions representing the horizontal plane, and one *(t)* temporal dimension. The spatiotemporal helix comprises a central spine and annotated prongs. More specifically:

- the central *spine* models the spatiotemporal trajectory described by the object's center as it moves during a time interval, while
- the protruding *prongs* express deformation (expansion or collapse) of the object's outline at a specific time instance.

As a spatiotemporal trajectory, a *spine* is a sequence of (x,y,t) coordinates. It can be expressed in a concise manner as a sequence of spatiotemporal nodes $S(n^1,...n^n)$. These nodes correspond to breakpoints along this trajectory, namely points where the object accelerated/decelerated and/or changed its orientation. Accordingly, *each node n^i is modeled as $n^i(x,y,t,q)$*, where:

- (x,y,t) are the spatiotemporal coordinates of the node, and
- q is a qualifier *classifying* the *node* as an *acceleration* (q^a), *deceleration* (q^d), or *rotation* (q^r) node.

Similarly, each prong is a model of the local expansion or collapse of the outline at the specific temporal instance when this event is detected, and is a horizontal arrow pointing away from (expansion) or towards (collapse) the spine. It is modeled as $p^i(t,r,a_1,a_2)$ where:

- t is the corresponding temporal instance (intersection of the prong and the spine in Fig. 1),
- r is the *magnitude* of this *outline* modification, expressed as a percentage of the local diameter,
- a_1, a_2 is the *range* of azimuths where this modification occurs; with each azimuth measured as a left-handle angle from the North (y) axis.

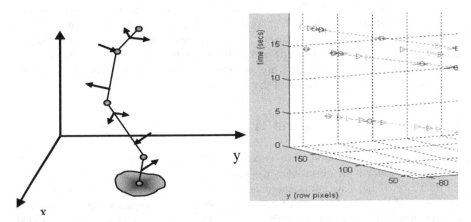

Fig. 1. The spatiotemporal helix model (left) and actual helixes generated for the video feed of fig. 2 (zoomed area of actual helix graph)

While Fig. 1 is a schematic diagram of the helix, its database representation is a sequence of n node and m prong records:

$$Helix^{objid}_{t1,t2} = (node^1,...node^n; prong^1,...prong^m)$$

Where node and prong records are as described above, *objid* is the identifier of the corresponding object and t_1, t_2 are the start and end instances of the time interval to which the helix refers.

This is visualized in Fig. 2 where we see a frame from a video feed captured by a sensor on top of a building observing a road segment with cars driving by, and pedestrians walking on the walkway. In the middle of Fig. 2 we see delineated car

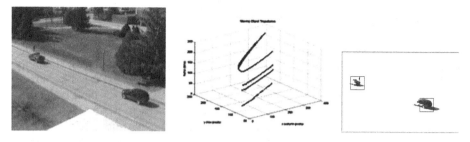

Fig. 2. A video frame (left), delineated trajectories (middle), and tracked objects (right)

trajectories. The 3-D space of Fig. 2(middle) is spatiotemporal, defined similarly to the convention used in Fig. 1. In this 3D space, car movements correspond to continuous spatiotemporal trajectories like the ones shown in Fig. 2(middle). While for this particular feed car trajectories are rather smooth, they can become quite complicated over broader areas, as the observed cars engage in more complex movements (e.g. accelerating /breaking, turning, stopping). While deformation is redundant when dealing with single objects, it becomes extremely important when describing groups of objects, as it communicates the closeness (and its variations) of members of a group.

Fig. 3. Visualization of helix query results in a GIS layer (left), or as video annotation (right) (courtesy of Milcord LLC, www.milcord.com)

Thus by tracking objects in video we can generate the corresponding helixes, which can be stored in any spatial database, which can then support various types of queries, like location-based queries (e.g, identifying instances when an object entered a region of interest) or time-based queries (e.g. identifying all objects that were moving during a specific time interval). The results of such queries can be displayed in a GIS, or overlapped in the corresponding video frames. Figure 3 shows an example of such a query from a prototype, with helix nodes (dots) identified in an area of interest (the light blue GIS layer) over a specific period of time. On the right hand side of the same figure we can see the result of a query visualized as annotated video.

4   Spatiotemporal Similarity Assessment

Spatiotemporal similarity assessment using helix information allows us to compare the behavior of the objects to which these helixes correspond. The simplest similarity measure is any L_p-norm function (e.g., the Euclidean distance), and it has been used for similarity assessment in trajectory/time-series datasets. However most real world phenomena may evolve at varying rates. Therefore there is a need for similarity/distance measures that allow time warping (i.e. allowing trajectories to stretch/shrink in the time domain). Two similarity assessment approaches appear most promising in addressing this goal:

- a higher-dimension *extension* of *Dynamic Time Warping* (DTW) [Berndt & Clifford, 1994] for helix comparison. and
- the Longest *Common SubSequence* (LCSS) approach to the same problem.

The difference between them is that LCSS allows the exclusion of certain sub-segments from the comparison of two longer segments, thus potentially leading to more robustness in terms of noise. LCSS is a variation of the edit distance [Levenshtein, 1966]. The length of the resulting subsequence can be used as the similarity measure [Agrawal et al., 1995, Das et al., 1997]. This function can be computed efficiently using dynamic programming. A visualization is offered in Fig. 4, with the comparison of two sequences.

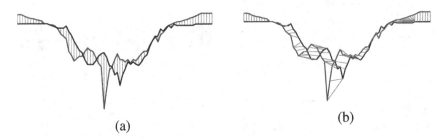

(a)

(b)

Fig. 4. (a) The Euclidean distance (L_2-norm); and (b) The DTW distance

4.1   Dynamic Time Warping

Most real-world phenomena can evolve at varying rates. For that reason – even though the vast majority of research on trajectory/time-series data mining has focused on the Euclidean distance – for virtually all real-world systems there is a need for similarity/distance measures that allow time warping (that is, elastic matching on the time domain). For example, in molecular biology it is well understood that functionally related genes will express themselves in similar ways, but possibly at different rates [Aach, 2001; Bar-Joseph, 2002]. For that reason DTW has been used extensively.

Let A and B be 2-dimensional trajectories of lengths n and m respectively. Also, let $Head(A)=((a_{x,1},a_{y,1}),...,(a_{x,n-1},a_{y,n-1}))$ be the first n-1 points of A.

Definition: The DTW distance between A and B is:

$$DTW(A,B) = L_p((a_{x,n}, a_{y,n}), (b_{x,m}, b_{y,m})) + \min \begin{bmatrix} DTW(Head(A), Head(B)) \\ DTW(Head(A), B) \\ DTW(A, Head(B)) \end{bmatrix}$$

The computation of DTW utilizes a dynamic programming technique. If the possible allowable matching in time is constrained within at most distance δ (the warping length) from each point, the computational cost of the algorithm is in the order of $O(\delta(n+m))$. This function gives a *measure of the distance* between two trajectories, as well. Figure 2(b) shows an example of DTW between two 1-dimensional trajectories.

The major drawback of this metric is that its efficiency deteriorates for noisy data. The algorithm matches individually all the points of a trajectory one by one. Thus, it also matches the outliers distorting the true distance. Another problem is that it is not suitable for use with most typical indexing techniques, since it violates the triangular inequality. Yet one more weakness is that it suffers from excessive computational cost for large warping lengths. Nevertheless, restricting the allowed warping length substantially speeds up the computation of DTW and yields other benefits as well (Vlachos et al., 2003).

4.2 Longest Common Subsequence

An alternative similarity measure is LCSS, which is a variation of the edit distance [Levenshtein, 1966]. The basic idea is to match two trajectories by allowing them to stretch on the time dimension without rearranging the order of the elements but allowing some of them to remain *unmatched*. The length of the resulting subsequence can be used as the similarity measure [Agrawal, et al., 1995; Bollobas, 1997; Bozkaya, 1997; Das et al., 1997]. Let A and B be 2-dimensional trajectories of lengths n and m respectively.

Definition: Given integers δ and ε we define the *LCSS* distance between A and B as:

$$LCSS_{\delta,\varepsilon}(A,B) = \begin{cases} 0, & \text{if } A \text{ or } B \text{ is empty} \\ 1 + LCSS_{\delta,\varepsilon}(Head(A), Head(B)), \\ \quad \text{if } |a_{x,n} - b_{x,m}| < \varepsilon \text{ and } |a_{y,n} - b_{y,m}| < \varepsilon \\ \quad \text{and } |n - m| \le \delta \\ max(LCSS_{\delta,\varepsilon}(Head(A), B), LCSS_{\delta,\varepsilon}(A, Head(B))), \\ \quad \text{otherwise} \end{cases}$$

The warping length δ controls the flexibility of matching in time and constant ε is the matching threshold in space. Contrary to the other two distance measures discussed so far, LCSS gives a *measure of the similarity* between two trajectories and not the distance. This function can be computed efficiently by dynamic programming and has complexity in the order of $O(\delta(n+m))$, if only a matching window δ in time is allowed [Das et al., 1997].

The value of LCSS is unbounded and depends on the length of the compared trajectories. Thus, it needs to be normalized in order to support trajectories of variable length. The distance derived from the LCSS similarity can be defined as follows:

Definition: The normalized distance $D_{\delta\varepsilon}$ expressed in terms of the LCSS similarity between A and B is given by:

$$D_{\delta,\varepsilon}(A,B) = 1 - \frac{LCSS_{\delta,\varepsilon}(A,B)}{\min(n,m)}$$

Even though LCSS presents similar advantages to DTW, it does not share its unstable performance in the presence of outliers. Nevertheless, this similarity measure is non-metric; hence, it cannot be used directly in combination with most typical indexing schemes.

4.3 Helix Pose Normalization

Helix comparison proceeds in a hierarchical manner, making progressive use of a decomposition of helix data. More specifically, spine and prong helix information can be analyzed to identify extreme events in them and generate a reduced description of helixes as sequences of these extreme events ($E=\{e_1, e_2...e_m\}$) in the behavior of the object. These short descriptions can be compared using either LCSS or DTW distance measures, to immediately eliminate poor candidates and accelerate our similarity comparison. Candidates that pass this initial test proceed for more detailed comparison through the introduction in the description of additional helix information (helix values that were not included in the first analysis). This process is repeated until all information is included, or all matching candidates are eliminated. Thus we introduce the notion of spatiotemporal decomposition into similarity assessment, with the potential of substantial computational gains.

This comparison scheme can benefit by the introduction of a helix normalization process, to make similarity assessment invariant to translation, rotation, and scale. Pose normalization transforms trajectories into a canonical coordinate frame that is entirely defined in terms of the helix data, thus eliminating the need for local invariant shape signatures that tend to become unstable in the presence of noise. This requires applying translation, rotation, and scale in 3D.

Helix pose normalization is closely related to the pose normalization problem in 3D object retrieval, where object similarity is estimated. [Elad et al., 2000] have proposed a normalization scheme that is based on Singular Value Decomposition (SVD) using second order object moments. Similarly to this approach, [Vranic et al., 2001] introduced the continuous Principal Component Analysis (cPCA) that ensures translation, rotation, reflection and scaling invariance. One of the key challenges in pose normalization is resolving the axis sign ambiguity, which is introduced since the SVD axis set is determined up to a sign. This requires object flipping, which occurs according to a predefined criterion, such as point counting or second order moments. Although several approaches have been developed to resolve the sign ambiguity, they tend to be inappropriate for spatiotemporal data as they are incapable of incorporating temporal aspects.

A force field approach works well for pose normalization, bypassing the above issues. It proceeds by estimating the amount of effort required to move a mass $m(t)$ that varies in time along the spine of the helix in the presence of a set of fixed masses, M, along the x, y and z axis (Fig. 5, left). In this configuration a spatiotemporal gravitational force field exists between M and $m(t)$. The notion of time flow as well as the spatial dimension is therefore represented here by the fact that $m(t)$ changes both its mass and location with time (Fig. 5, right). In this process helix flopping is controlled by the physical quantities of the force field (F_g) and the work (W):

$$\vec{F}_g(t) = G\frac{Mm(t)}{r(t)^2} \qquad W = \sum_{t=t_1}^{t_2} \left\langle \vec{F}_g, \vec{ds}(t) \right\rangle$$

Where $ds(t)$ is the normalized arc length along the helix, t_1 and t_2 are the first and last timestamp along the helix, and $<\cdot,>$ is the dot product.

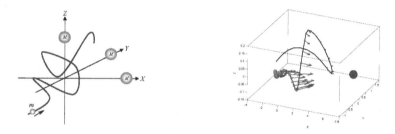

Fig. 5. Left: Placement of masses M along the x,y,z axes. **Right:** Example spatiotemporal force field.

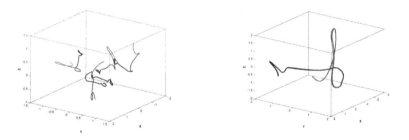

Fig. 6. Left: a set of identical trajectories that were subjected to random translation, rotation, and scale. **Right:** the same set of helixes after the normalization process.

Fig. 6 shows some experimental results, whereby a set of identical trajectories was subjected to random distortions (translation, rotation, scaling) as shown the left. These distortions were fully recovered using the force field approach. This ability to bypass such distortions is quite essential when attempting to compare activities that may take span different spatial and temporal ranges, yet may be inherently similar.

Through similarity assessment we can identify objects that act similarly (e.g. four vehicles moving in a similar manner, one behind the other), and group them together in composite objects (a convoy in this case). This composite object will have its own helix representation: its spine will represent the movement of the center of this convoy, and prongs will represent instances where the individual objects are grouped closer together, or move further apart.

5   Indexing Spatiotemporal Objects

If an appropriate similarity/distance measure can be decided based on a specific application domain, similarity queries can be answered in a straightforward way using exhaustive search – that is, computing the similarity/distance function between the query and every trajectory in the database. However, to speed up the computation it is important to avoid examining trajectories that are very distant to the query. This can be accomplished by discovering a close match as early as possible during the search, which can be used as a pruning threshold. A fast pre-filtering step can be employed that eliminates the majority of distant matches so that the costly, but accurate, similarity functions are executed only for a small subset of qualifying trajectories [Keogh, 2002; Yi, 1998].

Various indexing schemes can be used to aid in the pre-filtering step. For example, the trajectories can be approximated using Minimum Bounding Regions (MBR) and then stored in a multi-dimensional index structure [Hadjieleftheriou, 2002]. Alternatively, they can be represented as high-dimensional points and then stored in an index after using any dimensionality reduction technique [Agrawal, 1993].

Query response times can greatly benefit by the use of faster upper/lower-bounding functions of the actual similarity/distance measures (these are functions that consistently over/underestimate the true distance). These functions should guarantee no false dismissals. Intuitively, given an actual distance D of some trajectory from a query, one should be able to prune other trajectories for which the less-expensive lower-bounds D_{LB} have already been computed and for which $D_{LB} > D$. If, in addition, upper-bounds D_{UB} of the distance functions can be computed, any trajectory for which D_{LB} is larger than the minimum computed D_{UB}, can also be pruned; its actual distance from the query is definitely larger than D_{UB}. The inverse is true for *similarity* functions like LCSS.

5.1   Lower-Bounding the DTW

Most lower-bounding functions for DTW were originally defined for 1-dimensional time-series [Kim 2001; Yi et al., 1998]. A more robust lower-bounding technique was proposed by [Keogh, 2002]. Consider for simplicity a 1-dimensional time-series $A=(a_{x,1},...,a_{x,n})$. We would like to perform a very fast DTW match with warping length δ between trajectory A and query $Q(q_{x,1},...,q_{x,n})$. Suppose that we replicate each point $q_{x,i}$ for δ time instants before and after time i. The surface that includes all these points defines the area of possible matching in time, between the trajectory and the query. Everything outside this area should not be matched. We call the resulting area around query Q the *Minimum Bounding Envelope* (MBE) of Q.

The notion of the bounding envelope can be extended for more dimensions; for example, the $MBE_\delta(Q)$ for a 2-dimensional query trajectory $Q=((q_{x,1},q_{y,1}),...(q_{x,n},q_{y,n}))$ covers the area between the following trajectories:

$$EnvLow \leq MBE_\delta(Q) \leq EnvHigh$$

where for dimension d at position i:

$$EnvHigh_{d,i} = \max(q_{d,j}), \qquad |i - j| \leq \delta$$
$$EnvLow_{d,i} = \min(q_{d,j}), \qquad |i - j| \leq \delta$$

LB-Keogh works as follows: First, the MBE(Q) of query Q is constructed. Then, the distance of MBE(Q) from all other trajectories is evaluated. For trajectories Q and A (assuming dimensionality D), the distance between A and $MBE(Q)$ is:

$$DTW(MBE(Q), A) = \sqrt{\sum_{d=1}^{D} \sum_{i=1}^{n} \begin{cases} (a_{d,i} - EnvHigh_{d,i})^2, & \text{if } a_{x,i} > EnvHigh_{d,i} \\ (a_{d,i} - EnvLow_{d,i})^2, & \text{if } a_{x,i} < EnvLow_{d,i} \\ 0, & \text{otherwise} \end{cases}}$$

This function is the squared sum of the Euclidean distance between any part of A not falling within the envelope of Q and the nearest (orthogonal) edge of the envelope of Q, as depicted in Figure 7.

One can prove that LB-Keogh *lower-bounds* the actual time warping distance. A proof for the 1-dimensional case was given by Keogh (2002). This measure has since been used extensively in literature [Zhu, 2003; Vlachos et al., 2003]. Recently, in [Vlachos, 2004] we have shown that we can further improve the quality of the lower bound by computing a bounding box approximation of the trajectories, and performing a DTW computation on the approximations (Figure 7).

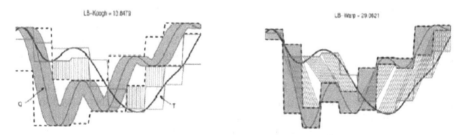

Fig. 7. A comparison of LB_Keogh (left) and LB_Warp (right)

5.2 Upper-Bounding the DTW

When performing a very fast LCSS match with warping length δ and within space ε between trajectory A and query $Q(q_{x,1},...,q_{x,n})$, the query MBE in this case should be extended within ε in space, above and below, (Figure 8).

The LCSS similarity between the envelope of Q and a trajectory A is defined as:

$$LCSS(MBE(Q), A) = \sum_{i=1}^{n} \begin{cases} 1, & \text{if } A[i] \text{ within envelope} \\ 0, & \text{otherwise} \end{cases}$$

This value represents an *upper-bound* for the *similarity* of Q and A. We can use the MBE (Q) to compute a *lower-bound* on the *distance* between trajectories as well: for any two trajectories Q and A the following holds:

$$D_{\delta\varepsilon}(MBE(Q),A) \leq D_{\delta\varepsilon}(Q,A)$$

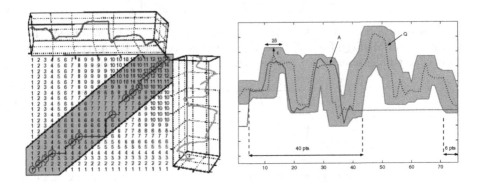

Fig. 8. (Left) Execution of the dynamic programming algorithm for LCSS. The warping length is indicated by the gray area (δ =6). (Right) The MBE within δ in time and ε in space of a trajectory. Everything that lies outside this envelope should not be matched.

5.3 Using Upper/Lower-Bounds for Quick Trajectory Pruning

In the two previous subsections we described techniques to upper/lower-bound the similarity/distance measures between two trajectories. According to the GEMINI framework [Agrawal et al., 1993], the approximated distance functions can be used to create an index that guarantees no false dismissals. However, the described upper/lower-bounds need to be computed using the raw trajectory data. A recent robust multi-dimensional indexing scheme for trajectories [Vlachos, 2003] can be used with any similarity/distance measures discussed so far and that does not need to access the actual trajectory data but only their compact approximations.

This scheme is based on the following principles. First, the trajectories are approximated using a large number of multi-dimensional MBRs which are then stored in an R-tree. For a given query Q, MBE(Q) is computed and decomposed into smaller MBRs. The resulting query MBRs are probed as range searches in the R-tree to discover which trajectory approximations intersect with Q and could be potential candidates. Trajectories that are very distant are never accessed and, thus, they are instantly pruned. Access to the raw trajectory data is restricted to only a few candidates.

This index is very compact since it stores only the substantially reduced in size trajectory approximations, and its construction time scales well with the trajectory length and the dataset cardinality. Therefore, this method can be utilized for massive data mining tasks. One of the significant advantages of this approach is its generality and flexibility. The user is given the ability to pose queries of variable warping length without the need to reconstruct the index. By adjusting the width of the bounding envelope on the query, the proposed method can support L_p-norm and constrained or full warping. Also, the user can choose between faster retrieval with approximate solutions, or exact answers on the expense of prolonged execution time.

6 Distributed Storage of Spatiotemporal Objects and In-Network Data Analysis

In this section we describe how we can collect and manage the sensor data in a distributed manner and how we can configure and reconfigure the sensor network system so that the capture and tracking of multiple objects can be optimized. Our approach is based on the integration of the tools for collecting, storing, managing and analyzing object trajectories with the protocols that manage the tracking of objects in the system. Efficient distributed indexing techniques for storing information about tracks, objects, and patterns, in the sensor network, can assist in the automatic classification of new object tracks. The problem we have to address in this setting is how to store, combine and analyze such data (a) to achieve real-time classification of new object tracks, while (b) minimizing the communication costs. For a concrete example, let us consider again the setting where each node stores a part of the different objects' trajectories, and we want to identify the objects that are most similar to a Query trajectory Q. The query trajectory may be an actual object (in this case we are interested into other similar objects), or it can be a composite trajectory, designed by the user to incorporate a set of characteristics, and the user is interested in finding the observed objects that best conform to these characteristics. To efficiently execute such a query we need both efficient index structures for storing local data in a node, and sophisticated algorithms for transmitting only the necessary data over the network. Since we have to consider the entire object if we want to find how similar it is to the query, we can avoid sending all the data over the network only if we can use approximation techniques to identify which objects are likely to be the results of our query.

6.1 The Distributed Most-Similar Trajectory Retrieval Problem

Let G denote a sensor network in a geographical area. Without loss of generality, we assume that the points in G are logically organized into cells. Each cell contains an access point (AP) that is assumed to be in communication range from every point in its cell.

Although the coordinate space is assumed to be partitioned in square cells, other geometric shapes such as variable size rectangles or Voronoi polygons are similarly applicable. This partitioning of the coordinate space simply denotes that in our setting, G is covered by a set of APs. Now let $\{A1,A2,...,Am\}$ denote a set of m objects moving in G. At each discrete time instance, object Ai ("i£m) generates a spatio-temporal record $r=\{Ai,ti,xi,yi\}$, where ti denotes the timestamp on which the record was generated, and (xi,yi) the coordinates of Ai at ti. The record r is then stored locally at the closest AP for l discrete time moments after which it is discarded. Therefore at any given point every access point AP maintains locally the records of the last l time moments.

The Distributed Most-Similar Trajectory Retrieval problem is defined as follows: Given a trajectory Q and a number K, retrieve the K trajectories which are the most similar to Q.

Example. Assume that some monitoring area G, is segmented into 4 cells, and that each cell is monitored by an access point (see Figure 9). A trajectory can be conceptually thought of as a continuous sequence Ai=((ax:1,y:1),...,(ax:l,y:l)) (i£m), while physically it is spatially fragmented across several cells. Similarly, the spatio-temporal query is also represented as: Q=((qx:1,y:1),...,(qx:l,y:l)) but this sequence is not spatially fragmented. The objective is to find the two trajectories that are most similar to the query Q (i.e. A1 and A2 for Fig. 9).

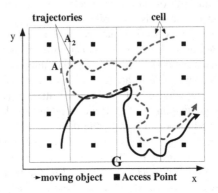

Fig. 9. Assuming a set of sensors, each storing the partial trajectory of an object, we want to develop efficient techniques that can store, retrieve, and analyze such data in-situ. The in-situ requirement allows us to minimize the impact of communication bottlenecks and improve the performance and the reliability of the system.

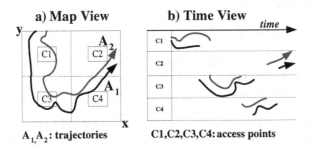

Fig. 10. Object trajectory representation in the original space (a) and in the local sensor space (b). For simplicity the time is not shown in (a), while (b) shows the projection of the trajectories to one dimension only.

The development of efficient techniques for evaluating queries such as the Most-Similar Trajectory Retrieval allows us to identify and group together similarly acting objects, manage the members of the group, and offer fault-tolerance guarantees under the probability of object (e.g. UAV or static sensor) failures or temporary network disconnections (i.e., short periods in which individual sensors are not able to communicate with the rest of the network). The sensor network can be configured to

C1 id,lb,ub	C2 id,lb,ub	C3 id,lb,ub	METADATA id,lb,ub
A2,3,6	A4,4,5	A4,1,3	A4,10,18
A0,4,8	A2,5,6	A0,6,10	A2,13,19
A4,5,10	A0,5,7	A2,5,7	A0,15,25
A7,7,9	A3,5,6	A9,6,7	A3,20,27
A3,8,11	A9,8,10	A3,7,10	A9,22,26
A9,8,9	A7,12,13	A7,11,13	A7,30,35
....	....	....	....

m (vertical, left)   *n* (horizontal, bottom)

1) Star   2) Hierarchy

Distributed Topologies

Fig. 11. An example of sensor networks with 3 sensors. Given a query trajectory Q, each sensor computes bounds on the similarity between the query Q and the parts of the trajectories of the objects that are locally stored at the sensor. Distributed top-K algorithms can be used then to find the trajectories that, given the bounds, are the most likely to be the most similar to Q. These trajectories are then retrieved by the query node, and their similarity to the query is computed exactly.

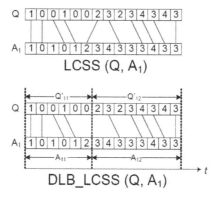

Fig. 12. Computing the longest common subsequence between a query trajectory Q and a trajectory A_1 that is stored in two sensors (A_{11}, A_{12}) in a distributed fashion (bottom figure) produces a lower bound on the actual longest common subsequence of Q and A_1 (top figure).

take specific roles or functionalities in managing the object groups. The roles can be assigned based on varying sensor node properties (such as available resources or location) or object tracking needs (e.g., similarly acting objects can be tracked by multiple sensor nodes for fault tolerance).

In the Distributed Top-K Trajectory Retrieval problem, the similarity query Q is initiated by some querying node QN, which disseminates Q to all the cells that intersect the query Q [Zeinalipour-Yazti et al., 2006]. We call the intersecting cells candidate cells. Upon receiving Q, each candidate cell executes locally a lower bounding matching function (LowerM) and an upper bounding function (UpperM) on all its local subsequences. This yields 2m local distance computations to Q by each cell (one for each bound). The conceptual array of lower (LB) and upper bounds (UB) for an example scenario of three nodes (C1,C2,C3) is illustrated in Figure 11. We will refer to the sum of bounds from all cells as METADATA and to the actual subsequence trajectories stored locally by each cell as DATA. Obviously, DATA is orders of magnitudes more expensive than METADATA to be transferred towards QN. Therefore we want to intelligently exploit METADATA to identify the subset of DATA that produces the K highest ranked answers. Figure 11 illustrates two typical topologies between cells: star and hierarchy. Our proposed algorithms are equivalently applicable to both of them although we use a star topology here to simplify our description.

Using this approach we can prune down the number of trajectories that need to transmitted to the query node, in order to find the most similar ones to the query. If we did not utilize the bounds, on order to find the K trajectories that are most similar to a query trajectory Q, the query QN can fetch all the DATA and then perform a centralized similarity computation using the FullM(Q,Ai) method, which is one of the LCSS, DTW or other Lp-Norm distance measures. The centralized is clearly more expensive in terms of data transfer and delay. Figure 12 shows that computing the similarity in a distributed fashion (in the example of figure 12 the similarity measure is the Longest Common SubSequence) only gives a lower bound to the actual similarity; consequently some of the trajectories still have to be retrieved by the query node so that the true most similar trajectory can be determined. Figure 13 shows that enveloping techniques can be employed to further speed up the computation of the bounds on the similarity.

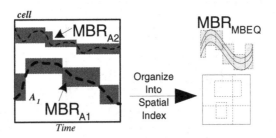

Fig. 13. A promising approach for storing and indexing trajectories is to compress the trajectories by encapsulating them with a small number of minimum bounding rectangles (MBRs). In addition to good accuracy, a great advantage of this approach is that we can leverage the significant body of research by the database community on low dimensionality indexing structures, such as R-trees and their variants.

6.2 Real-Time Object Tracking

The sensor network setting we consider consists of stationary sensors and mobile sensors (for example, UAV- mounted, or building mounted with the ability to rotate, pan and zoom), and the surveillance area is tessellated into a grid. We assume that each sensor knows its location, and the geographic boundaries of the cells. In an ideal situation a single sensor can be responsible for each grid cell. However, due to failures and energy expenditures, this scenario is quite unlikely [Halkidi et al., 2007]. Instead, we would typically encounter situations where multiple sensors may be responsible for a single cell, or one mobile sensor may be responsible for more than a single cell.

Thus we face the challenge of managing the network so as to promptly detect a moving object as it approaches, and generate and forward messages to notify appropriate sensors in an energy-efficient way.

To address this issue we introduce a *novel framework for sensor management.* Considering our application focus (tracking moving objects) the main issue we address is handing-off tracking duties from one sensor to the rest of the network as an object is exiting this sensor's field of view. In order to visualize our approach let's consider Fig. 14. We see the tessellation of space, and the location of the sensors, indicated by the white dots. For each cell there is a single sensor that is responsible for managing the cell, indicated in the figure as blue dots. Such a sensor is called the leader, and in our mechanism is selected periodically at random among the sensors in the cell. This is done to distribute to all the sensors the impact that the work that the leader has to perform has to the individual sensor resources. The goal of our framework is to enable sensors to collaborate and pass information among themselves. The distinguishing characteristic of our framework is that it involves different algorithms for *tracking* and *warning*. This approach can be summarized as follows:

1. **Tracking Loop:** Tracking loop performs in-cell analysis, and is executed at each sensor. For all objects tracked in this cell we can generate their trajectories. Using these trajectories and online techniques such as Kalman filtering, and the predictive models discussed earlier, we will predict the expected locations of the target. This is performed for all objects tracked within this cell. The predicted locations of all these objects can also be used to control the movements of the mobile sensors that may track the object.

2. **Warning Loop:** Warning loop addresses cell crossings by a target. This is visualized in Fig. 14 by the red triangle (representing the target) that is ready to move outside the cell where it currently resides. In this situation the cell leader will alert its neighboring leaders, so that they can become aware that a new object is about to enter their cell. When this is done, the first cell control station will also forward the tracked object information, so as to best support prediction and tracking in the next cells. This is in essence a *hand-off* technique, with the control stations exchanging information and alerting each other.

Example: In Fig. 14 all neighboring cells have active sensors operating in them; however, as we mentioned, this may not be the case. When there is no object close to a cell, the sensors of that cell may be offline (to preserve energy), in which case the leader must use a wake up protocol to wake some of them up. It may also be the case that no sensors exist in this cell at all, but it is covered by the dispatch of mobile sensors. In such a case, the tracked information will be forwarded to a central control unit that will maintain a log of objects that entered unmonitored cells. If this log indicates that a substantial number of objects enters unmonitored cells, the central control unit may suggest the reconfiguration of the network so that some of the currently unmonitored cells are also covered (for example, moving a mobile sensor from one cell to another, or changing the sleep cycle of the sensors in the cell).

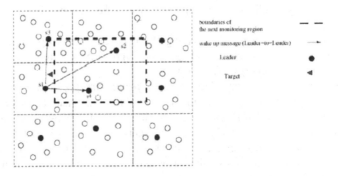

Fig. 14. Alert messages are forwarded from S_1 to the leaders of the neighboring cells (S_2, S_3 and S_4), which are intersected with the region that the target is expected to traverse

Such a system scales well with the number of mobile objects, or the number of concurrently tracked objects, to easily accept and use more sensors on-line, and to degrade gracefully when UAVs malfunction or leave an area. This innovative framework not only *controls* the placement of sensors, but also manages the storage of information within our network. As an object moves from one cell to another, its record follows it. This allows us to have immediate access to an object's complete information, including its trajectory before entering a specific location, or before it was acquired by a specific sensor.

7 Conclusions

In this paper we addressed the issue of mobility in geosensor networks. We presented key components of a genaral framework for this topic, considering mobility two levels, with mobile objects in the area monitored by a network of mobile sensors. More specifically, we presented methods to model the spatiotemporal variations in position and outline of an object in section 3. The spatiotemporal helix allows us to model the movement of a car, or the defomrations of a composite object over time, information that can be recorded in in a spatial database. The content of these records can the use of be analyzed to compare two or more moving objects, using the techniques presented in section 4, namely dynamic time warping and LCSS. Through pose normalization this matching can become independent of rotations, shifts and scalings, while the use of upper/lower-bounding functions can substantially improve query response time as described in section 5. Together these solutions allow us to move from captured spatiotemporal trajectories to comprehensive spatial databases describing their content, and similarity metrics to support complex analysis and identify similarly behaving objects. The approach we presented in Section 6 make suse of these capabilities to define a management plan for mobile sensors in highly active areas. Even though our inspiration stems from surveillance using distributed optical sensors (e.g. on-board unmanned aerial vehicles), most of the solutions presented in this paper are in principle applicable to a wide array of spatiotemporal datasets (e.g. collected using RFID tags or GPS-enabled cell phones).

Acknowledgments. This work was supported by the National Science Foundation through NSF Awards ITR 0121269, IIS 0330481, CNS 0627191, and IIS 0534781, and by the National Geospatial-Intelligence Agency through NURI award NMA 401-02-1-2008. We would also like to acknowledge the input of Dr. Arie Croitoru on the development of the approach outlined in section 4.3.

References

1. Aach, J., Church, G.: Aligning Gene Expression Time Series Data. Bioinformatics 17, 495–508 (2001)
2. Abdelzaher, T., Blum, B., Cao, Q., Evans, D., George, J., George, S., He, T., Luo, L., Son, S., Stoleru, R., Stankovic, J., Wood, A.: EnviroTrack: Towards an Environmental Computing Paradigm for Distributed Sensor Networks. In: Proc. Int. Conf. on Distributed Computing Systems (ICDCS 2004), pp. 582–589 (2004)
3. Agouris, P., Stefanidis, A.: Efficient Summarization of SpatioTemporal Events. Communications of the ACM 46(1), 65–66 (2003)
4. Agrawal, R., Faloutsos, C., Swami, A.: Efficient Similarity Search in Sequence Databases. In: Lomet, D.B. (ed.) FODO 1993. LNCS, vol. 730, pp. 69–84. Springer, Heidelberg (1993)
5. Agrawal, R., Lin, K., Sawhney, H.S., Shim, K.: Fast Similarity Search in the Presence of Noise, Scaling and Translation in Time-Series Databases. In: Proc. of the VLDB, pp. 490–501 (1995)
6. Ailamaki, A., Faloutsos, C., Fischbeck, P., Small, M., VanBriesen, J.: An Environmental Sensor Network to Determine Drinking Water Quality and Security. SIGMOD Record 32(4), 47–52 (2003)
7. Altenis, S., Jensen, C.S.: Indexing of Moving Objects for Location-Based Services, Department of Computer Science, Aalborg University (2002)
8. Aslam, J., Butler, Z., Constantin, F., Crespi, V., Cybenko, G., Rus, D.: Tracking a Moving Object with a Binary Sensor Network. In: Proc. Int. Conf. on Embedded Networked Sensor Systems (SenSys 2003), pp. 150–161 (2003)
9. Bandyopadhyay, S., Coyle, E.J.: An Energy Efficient Hierarchical Clustering Algorithm for Wireless Sensor Networks. In: IEEE INFOCOM 2003, pp. 1713–1723 (2003)
10. Bar-Joseph, Z.G., Gerber, D., Gifford, T.: A New Approach to Analyzing Gene Expression Time Series Data. In: Proc. Annual Int. Conf. on Research in Computational Molecular Biology, pp. 39–48 (2002)
11. Berndt, D., Clifford, J.: Using Dynamic Time Warping to Find Patterns in Time Series. In: Proc. of KDD Workshop, pp. 359–370 (1994)
12. Beymer, D., McLaughlan, P., Coifman, B., Malik, J.: A Real-Time Computer Vision System for Measuring Traffic Parameters. In: Computer Vision Pattern Recognition (CVPR 1997), pp. 495–500 (1997)
13. Bolles, R., Nevatia, R.: A Hierarchical Video Event Ontology in Owl, Pacific Northwest National Laboratory, Technical Peport PNNL-14981 (2004)
14. Bollobas, B., Das, G., Gunopulos, D., Mannila, H.: Time-Series Similarity Problems and Well-Separated Geometric Sets. In: Proc. of the 13th SCG, pp. 243–307 (1997)
15. Bozkaya, T., Yazdani, N., Ozsoyoglu, M.: Matching and Indexing Sequences of Different Lengths. In: Proc. of the CIKM, pp. 128–135 (1997)
16. Brooks, R., Ramanathan, P., Sayeed, A.: Distributed Target Classification and Tracking in Sensor Networks. Proceedings of the IEEE 91(8), 1163–1171 (2003)

17. Carney, D., Cetintemel, U., Cherniack, M., Convey, C., Lee, S., Seidman, G., Stonebraker, M., Tatbul, N., Zdonik, S.: Monitoring Streams - A New Class of Data Management Applications. In: Proc. VLDB (2002)
18. Cerpa, A., Elson, J., Hamilton, M., Zhao, J., Estrin, D., Girod, L.: Habitat Monitoring: Application Driver for Wireless Communications Technology. In: Workshop on Data communication in Latin America and the Caribbean, San Jose, Costa Rica, pp. 20–41 (2001)
19. Chang, T.H., Gong, S., Ong, E.J.: Tracking Multiple People Under Occlusion Using Multiple Cameras. In: Proc. British Machine Vision Conf. (2000)
20. Chen, A., Muntz, R., Srivastava, M.: Smart Rooms. In: Cook, D., Das, S. (eds.) Smart Environments: Technology, Protocols and Applications, Wiley, Chichester (2004)
21. Collins, R., Lipton, A., Fujiyoshi, H., Kanade, T.: Algorithms for Cooperative Multisensor Surveillance. Proceedings of IEEE 89(10), 1456–1477 (2001)
22. Conner, S., Heidemann, J., Krishnamurthy, L., Wang, X., Yarvis, M.: Workplace Applications of Sensor Networks. In: Bulusu, N., Jha, S. (eds.) Wireless Sensor Networks: A Systems Perspective, Artech House, pp. 289–308 (2005)
23. Das, G., Gunopulos, D., Mannilaz, H.: Finding Similar Time Series. In: Komorowski, J., Żytkow, J.M. (eds.) PKDD 1997. LNCS, vol. 1263, pp. 88–100. Springer, Heidelberg (1997)
24. Dockstader, S., Tekalp, A.M.: Multiple Camera Fusion for Multi-Object Tracking. In: Proc. IEEE Workshop on Multi-Object Tracking (WOMOT 2001), pp. 95–100 (2001)
25. Elad, M., Tal, A., Ar, S.: Directed Search in a 3D Objects Database. Technical Report, HP Labs (2000)
26. Eltoukhy, H., Salama, K.: Multiple Camera Tracking, Stanford Image Sensors Group, Electrical Engineering Department, Stanford University (2001)
27. Faradjian, A., Gehrke, J., Bonnet, P.: GADT: A Probability Space ADT for Representing and Querying the Physical World. In: Int. Conf. on Data Engineering (ICDE 2002), pp. 201–206 (2002)
28. Gehrke, J., Korn, F., Srivastava, D.: On Computing Correlated Aggregates Over Continual Data Streams. In: ACM Int. Conf. on Management of Data (SIGMOD), pp. 13–24 (2001)
29. Hadjeleftheriou, M., Kollios, G., Tsotras, V., Gunopulos, D.: Efficient Indexing of Spatiotemporal Objects. In: Jensen, C.S., Jeffery, K.G., Pokorný, J., Šaltenis, S., Bertino, E., Böhm, K., Jarke, M. (eds.) EDBT 2002. LNCS, vol. 2287, pp. 251–268. Springer, Heidelberg (2002)
30. Halkidi, M., Papadopoulos, D., Kalogeraki, V., Gunopulos, D.: Resilient and Energy Efficient Tracking in Sensor Networks. Int. J. of Wireless and Mobile Computing (2007) (in press)
31. Huang, Q., Lu, C., Roman, G.: Mobicast: Just-in-Time Multicast for Sensor Networks under Spatiotemporal Constraints. In: Zhao, F., Guibas, L.J. (eds.) IPSN 2003. LNCS, vol. 2634, pp. 442–457. Springer, Heidelberg (2003a)
32. Huang, Q., Lu, C., Roman, G.: Spatiotemporal Multicast in Sensor Networks. In: Proc. Int. Conf. on Embedded Networked Sensor Systems (SenSys 2003), pp. 205–217 (2003b)
33. Javed, O., Khan, S., Rasheed, Z., Shah, M.: Camera Handoff: Tracking in Multiple Uncalibrated Stationary Cameras. In: Proc. Workshop on Human Motion, pp. 113–121 (2000)
34. Javed, O., Rasheed, Z., Shafique, K., Shah, M.: Tracking across Multiple Cameras with Disjoint Views. In: IEEE Int. Conf. on Computer Vision, vol. 2, pp. 952–957 (2003)
35. Jaynes, Ch.: Acquisition of a Predictive Markov Model using Object Tracking and Correspondence in Geospatial Video Surveillance Networks. In: Stefanidis, A., Nittel, S. (eds.) GeoSensor Networks, pp. 149–166 (2004)

36. Jaynes, C., Webb, S., Steele, R., Xiong, Q.: An Open-Development Environment for the Evaluation of Video Surveillance Systems. In: Proc. PETS, Copenhagen (2002)
37. Juang, P., Oki, H., Wang, Y., Martonosi, M., Peh, L., Rubenstein, D.: Energy-Efficient Computing for Wildlife Tracking: Design Tradeoffs and Early Experiences with ZebraNet. In: Proc. Intl. Conf. On Architectural Support for Programming Languages and Operating Systems (ASPLOS-X), San Jose, CA, pp. 96–107 (2002)
38. Keogh, E.: Exact Indexing of Dynamic Time Warping. In: Proc. VLDB, pp. 406–417 (2002)
39. Kim, S., Park, S., Chu, W.: An Index-Based Approach for Similarity Search supporting Time Warping in Large Sequence Databases. In: Proc. of the ICDE, pp. 607–614 (2001)
40. Levenshtein, V.: Binary Codes Capable of Correcting Deletions, Insertions, and Reversals. Soviet Physics-Doklady 10(10), 707–710 (1966)
41. Liu, J., Reich, J., Cheung, P., Zhao, F.: Distributed Group Management for Track Initiation and Maintenance in Target Localization Applications. In: Zhao, F., Guibas, L.J. (eds.) IPSN 2003. LNCS, vol. 2634, pp. 113–128. Springer, Heidelberg (2003)
42. Madden, S., Franklin, M., Hellerstein, J., Hong, W.: Tag: a Tiny Aggregation Service for ad-hoc Sensor Networks. In: OSDI 2002, pp. 131–146 (2002)
43. Mainwaring, A., Polastre, J., Szewczyk, R., Culler, D.: Wireless Sensor Networks for Habitat Monitoring, Technical Report IRB-TR-02-006, Intel Laboratory, UC Berkeley (2002)
44. Makris, D., Ellis, T.: Path Detection in Video Surveillance. Image and Vision Computing Journal 20(12), 895–903 (2002)
45. Motwani, R., Widom, J., Arasu, A., Babcock, B., Babu, S., Datar, M., Manku, G., Olston, C., Rosenstein, J., Varma, R.: Query Processing, Resource Management, and Approximation in a Data Stream. In: Proc. Conference on Innovative Data Systems Research (CIDR), pp. 245–256 (2003)
46. Nascimento, M., Pfoser, D., Theodoridis, Y.: Synthetic and Real Spatiotemporal Datasets. Data Engineering Bulletin 26(2), 26–32 (2003)
47. Needham, C.J., Boyle, R.D.: Performance Evaluation Metrics and Statistics for Positional Tracker Evaluation. In: Crowley, J.L., Piater, J.H., Vincze, M., Paletta, L. (eds.) ICVS 2003. LNCS, vol. 2626, pp. 278–289. Springer, Heidelberg (2003)
48. Nittel, S., Stefanidis, A.: GeoSensor Networks and Virtual GeoReality. In: Stefanidis, A., Nittel, S. (eds.) GeoSensor Networks, pp. 1–9. CRC Press, Boca Raton (2004)
49. Porikli, F.: Trajectory Distance Metric using Hidden Markov Model Based Representation. In: Pajdla, T., Matas, J(G.) (eds.) ECCV 2004. LNCS, vol. 3024, pp. 39–44. Springer, Heidelberg (2004)
50. Rhodes, B., Bomberger, N., Seibert, M., Waxman, A.: Maritime Situation Monitoring and Awareness using Learning Mechanisms. In: IEEE MILCOM 2005, pp. 646–652 (2005)
51. Srivastava, M., Muntz, R., Potkonjak, M.: Smart Kindergarten: Sensor-Based Wireless Networks for Smart Developmental Problem-Solving Environments. In: Proc. of ACM SIGMOBILE, pp. 132–138 (2001)
52. Stefanidis, A., Eickhorst, K., Agouris, P., Partsinevelos, P.: Modeling and Comparing Change using Spatiotemporal Helixes. In: Hoel, E., Rigaux, P. (eds.) ACM-GIS 2003, pp. 86–93. ACM Press, New York (2003)
53. Stauffer, C., Grimson, W.E.L.: Learning Patterns of Activity using Real-Time Tracking. IEEE Trans. on Pattern Analysis & Machine Intelligence 22(8), 747–757 (2000)
54. Stauffer, C., Tieu, K.: Automated Multi-Camera Planar Tracking through Correspondence Modeling. In: Proc. Computer Vision & Pattern Recognition, vol. I, pp. 259–266 (2003)

55. Vlachos, M., Hadjieleftheriou, M., Gunopulos, D., Keogh, E.: Indexing Multi- Dimensional Time-Series with Support for Multiple Distance Measures. In: Proc. SIGKDD, pp. 216–225 (2003)
56. Vlachos, M., Meek, C., Vagena, Z., Gunopulos, D.: Identifying Similarities, Periodicities and Bursts for Online Search Queries. In: Proc. ACM SIGMOD, pp. 131–142 (2004)
57. Vranic, D., Saupe, D.: Tools for 3D Object Retrieval: Karhunen-Loeve Transform and Spherical Harmonics. In: Proc. IEEE Work. on Multimedia Signal Processing, pp. 293–298 (2001)
58. Yand, H., Sikdar, B.: A Protocol for Tracking Mobile Targets using Sensor Networks. In: IEEE Int. Workshop on Sensor Networks Protocols and Applications, pp. 71–81 (2003)
59. Ye, F., Luo, H., Cheng, J., Lu, S., Zhang, L.: A Two-Tier Data Dissemination Model for Large-scale Wireless Sensor Networks. In: MOBICOM 2002, pp. 148–159 (2002)
60. Yi, B.-K., Jagadish, H.V., Faloutsos, C.: Efficient Retrieval of Similar Time Sequences under Time Warping. In: Proc. of the ICDE, pp. 201–208 (1998)
61. Zeinalipour-Yazti, D., Lin, S., Gunopulos, D.: Distributed Spatio-Temporal Similarity Search. In: Proc. of ACM CIKM, pp. 14–23 (2006)
62. Zhang, W., Cao, G.: Optimizing Tree Reconfiguration for Mobile Target tracking in Sensor Networks. In: IEEE INFOCOM 2004, vol. 4, pp. 2434–2445 (2004)
63. Zhu, H., Su, J., Ibarra, O.H.: Trajectory Queries and Octagons in Moving Object Databases. In: Proc. of ACM CIKM, pp. 413–421 (2002)

Continuous Spatiotemporal Trajectory Joins

Petko Bakalov and Vassilis J. Tsotras

Computer Science Department, University of California, Riverside
{pbakalov,tsotras}@cs.ucr.edu

Abstract. Given the plethora of GPS and location-based services, que- ries over trajectories have recently received much attention. In this paper we examine trajectory joins over streaming spatiotemporal data. Given a stream of spatiotemporal trajectories created by monitored moving objects, the outcome of a *Continuous Spatiotemporal Trajectory Join* (CSTJ) query is the set of objects in the stream, which have shown similar behavior over a query-specified time interval, relative to the current timestamp. We propose a novel indexing scheme for streaming spatiotemporal data and develop algorithms for CSTJ evaluation, which utilize the proposed indexing scheme and effectively reduce the computation cost and I/O operations. Finally, we present a thorough experimental evaluation of the proposed indexing structure and algorithms.

1 Introduction

The abundance of position locators and GPS devices enables creation of data management systems that monitor streaming spatiotemporal data, providing multiple services to the end users. The basic architecture of a monitoring system working with spatial streaming data consists of multiple tracing devices which continuously report their location thus forming a spatiotemporal stream. Such streams are collected to the base station (server) where users submit their queries and continuously receive query results based on the current state of the data. Unlike traditional snapshot queries that are evaluated only once continuous queries require continuous revaluation as the query result becomes obsolete and invalid with the change of information for the objects.

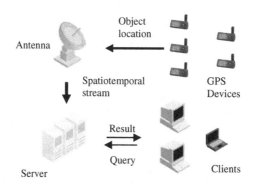

Fig. 1. A streaming spatiotemporal architecture

S. Nittel, A. Labrinidis, and A. Stefanidis (Eds.): GSN 2006, LNCS 4540, pp. 109–128, 2008.
© Springer-Verlag Berlin Heidelberg 2008

Recent research efforts have focused mainly on simple queries (i.e., having just a single spatial Range [7][23] or Nearest Neighbor queries [33] predicate). However in many real life monitoring queries there is need for more complex continuous spatial predicates. For example, users may be interested in discovering pairs of moving objects which follow similar movement pattern for specified period of time. Consider a security system inside a building which is tracing person movement. The security officer may want to continuously check for any security violations or suspicious activities over the stream of spatiotemporal data coming from the sensors inside the building. One suspicious activity for example can be "Identify the pairs of (security officers,visitors) that have followed each other in the last 10 minutes." since it can be sign for someone trying to study the security personal.

In this paper we address a novel query for streaming data, called a *trajectory join*, i.e., the problem of identifying all pairs of similar trajectories between two datasets. The trajectory similarity is defined by their spatial closeness (expressed in a condition by a spatial threshold ϵ around each trajectory) which should last at least for an interval with duration δt. So the trajectory join can be expressed as a complex query predicate involving both the spatial and temporal constraints over the object trajectories. In the definition of the trajectory join problem for a static data set scenario [2][3] the temporal constraint δt is an absolute one (For example "Between 2 a.m and 3 a.m"). The absolute temporal constraints however do not make sense in a continuous environment since the result for them never changes. More useful for continuous queries are the relative time constraints. A relative time constraint uses the current time instance as a reference point. (Example: between 2 and 3 hours ago). As the time passes, the value of the current timestamp changes which makes the relative time constraint slide along the the temporal axis.

Because of the constant reevaluation of the result and the use of relative time constraints instead of absolute ones the extension of the static join solutions in the continuous environment is not efficient. A trivial extension of the existing static algorithms to continuous version can be repetitive execution of the static algorithm every time when the result has to be refreshed. However this is very expensive as the query evaluation starts from the beginning every time when the result has to be refreshed.

The prevailing strategy for efficient continuous query evaluation is the incremental approach [34,33]. This approach implies that the query processor reuse as much as possible the current result and data structures for future iterations of the evaluation algorithm.

Nevertheless the CSTJ problem differs form all other continuous spatial predicates in that it also involves historical data from the stream.

In order to adopt efficiently the incremental approach for evaluation of the CSTJ queries we need an indexing structure for streaming data, which is able to:

- Answer queries about previous states of the spatiotemporal stream.
- Provide approximation for the object trajectory.
- Support the incremental approach for query evaluation.

To the best of our knowledge there is no indexing schema proposed, which has all these properties. In this paper we propose a novel indexing structure for spatial streams which is able to store information about previous states of the spatiotemporal stream and develop algorithms for CSTJ evaluation, which utilize this indexing structure.

2   Related Work

Many join and self-join algorithms have been designed and proposed in the past for different data types and more specifically for spatial data [5] [19] [21] [27] [13] [31] [10] [35] [25] [18] [20] [1] [8]. However, these algorithms are not applicable in the case of spatio-temporal trajectories because they are based on intersections between MBRs while spatiotemporal trajectory join conditions are much more complex with constraints in both the spatial and temporal domain.

Recent work in the area of spatiotemporal streams has led to multiple indexing techniques and processing algorithms. They can be divided generally in three groups.

In [7] [23] [26] [24] the use of a simple grid structure is proposed for indexing the location of the objects inside the spatiotemporal stream. Every single grid cell is associated with a list of the objects currently residing inside it. Clearly such approach is very efficient from a computational point of view since the maintenance of the index structure is straightforward. It can handle very effectively issues like frequent updates, high arrival raters, the infinite nature of the data and so on. However it can be used only for a queries focused on the current state of the stream.

Multiple algorithms have been proposed for answering range predicates with the grid based indexing solutions. Gedik and Liu [7] propose a distributed system for range queries called "mobieyes", and it is assumed that the moving clients can process and store information. The client receive information about the moving range query from the server and notifies the server when it is going to enter or leave this query region. Mokbel et al [23] implements SINA, a scalable shared execution and incremental evaluation system, where the execution of continuous range spatiotemporal queries is abstracted as a spatial join between the set of objects and the set of queries. Continuous evaluation of nearest neighbor queries have also received a lot of attention lately using grid structures [12] [34] [33] [24]. Koudas et al [12] propose DISC a technique for answering ϵ-approximate k nearest neighbor queries over a stream of multidimensional data. The returned k^{th} Nearest neighbor lies at most on distance $d + \epsilon$ from the query point where d is the distance between the actual k^{th} Nearest neighbor and the query point. Yu at al.[34] and Xiong at al. [33] propose similar approaches for answering continuous NN queries using different distances for pruning. Finally Mouratidis et al. [24] introduced conceptual partitioning which archives a better performance than the previous approaches by handling updates from objects which fall in vicinity of the query.

The second group of indexing methods uses different tree-like structures. There are structures based on B+-trees [17] [11], R trees [14] [15] and TPR-trees [32] [16] [29] [30]. The main objective is to improve the update performance of the indexing structure since it is the most frequent operation in streaming environment. In [14] the reduction of the update cost is done trough avoiding the updates for objects that do not move outside of their MBRs. Later in [15] this technique is generalized trough a bottom - up update strategy which uses different levels of reorganization during the updates and in this way avoids the expensive top-down updates. The minimization of the update time in [17] [11] is achieved trough the use of B+ trees, which have better update characteristics, instead of traditional multidimensional index structures like R-tree [9] [4]. This is achieved trough linearization of the representation of the moving objects locations using space filing curves like the Peano [6] or Z curve.

The last group [28,22] of query evaluation methods for streams tries to avoid the expensive maintenance of index structures over the data. These methods are based on the notion of "safe" regions, created around the data [28] or uncertainty regions around the query [22]. If the object doesn't leave its safe region no further processing is required. And the reverse - in [22] objects are considered only if they fall inside the query region or its uncertainty regions.

All these indexing structures, discussed so far, try to improve the performance by minimizing the update rate. To the best of our knowledge there has not been any approach to improve the index performance from point of view of the query evaluation strategy. Later in this paper we propose a novel indexing structure which has fast object update rate and is oriented towards the incremental evaluation (i.e. an approach that reuses the result from the previous step).

3   Problem Definition

Consider a system that continuously monitors the locations of a set of moving objects. Location updates arrive as a stream of tuples $S = \langle u_1, u_2, \ldots, u_l, \ldots \rangle$ where $u_i = \langle o_i, l_i, t_i \rangle$, and o_i is the object issuing the update while l_i is the new location of the object on the plane and t_i is the current time stamp. $l_i \in \mathbb{R}^d, o_i \in \mathbb{N}$ (for simplicity we can assume a two dimensional plane).

Trajectory. $T(o_i)$ of an object o_i in a stream S is a sequence of pairs $\{\langle l_1, t_1 \rangle, \ldots, \langle l_n, t_n \rangle\}$, where $l_i \in \mathbb{R}^d, t_i \in \mathbb{N}$. Let t_{now} denote the ever increasing current time instant ($t_{now} \in \mathbb{N}$). The definition of the CSTJ query follows:

Given trajectory sets r and s, the CSTJ query continuously returns all trajectory pairs $\langle T(o_{ri}), T(o_{si}) \rangle$ which have been spatially close (within threshold ϵ) for some time period δt ending at the current timestamp (i.e. the temporal constraint uses as a reference point the current timestamp). An example of such a relative time constraint is the restriction "in the last 30 minutes". In contrast, absolute time constraints (e.g. "between 2:30pm and 3:40pm") produce a query result that is static and does not change with time. In a continuous query environment, as the current time proceeds some objects will expire from the observed period while others will be introduced, thus continuously changing the join result.

Continuous Spatiotemporal Trajectory Join. Given two sets of moving objects o_r and o_s, a spatial threshold ϵ and a (relative) time period δt ($\delta t \in \mathbb{N}$), the CSTJ returns continuously the set of pairs $\langle o_{ri}, o_{sj} \rangle$ such that for every time instance t_i between t_{now} and $t_{now} - \delta t$ the spatial distance between the trajectories $T(o_{ri})$ and $T(o_{sj})$ is less than the threshold ϵ.

4   Evaluation Framework

The basic idea behind the evaluation algorithms for the static version of the problem [2][3] is to find a way to prune as many trajectory pair similarity evaluations as possible. There are two major elements needed for the efficient evaluation of a trajectory join, namely:

- First we need a compact object trajectory approximation. This requirement is necessary to make the index structure which stores the trajectory approximations small and thus fit into the main memory for efficient access. It is assumed that the raw spatiotemporal stream data is too large to be kept in the main memory and has to be stored on a secondary storage devices. We further assume that the raw trajectory data is stored in lists of data pages per trajectory where each data page has a pointer to the next one in the list.
- Second, we require an easy to compute lower bound distance function between the trajectory approximations.

Using trajectory approximations and lower bound distance functions we can prune a large number of the pairs from the Cartesian product between the object trajectory sets $T(o_r)$ and $T(o_s)$. Because we work with trajectory approximations instead of the actual (full) trajectory data a verification step is also needed, where the pairs of trajectories, not pruned away by the distance function are then verified to satisfy the join criteria using their actual trajectory data. The lower bound distance function defined in this paper guarantees that we may have only "false positives" in the verification step (i.e., some trajectory pairs not pruned away by the lower bound distance may still not satisfy the join criteria) but no "false negatives" (i.e., no join result is missed). To remove these false positives in the final result we need the extra verification step which access the raw trajectory data on the secondary storage device and verifies for each pair that it indeed satisfies the join criteria. Hence the total cost of a single evaluation iteration will comprise of two parts:

- The cost of computing the lower-bounding distances.
- The cost of executing the verification step.

For the continuous version of the problem there are also additional requirements. The trajectory approximation should be easy to compute and maintain. This requirement is needed since the approximation is created on the fly as the streaming data enters the server. For example the static approximation discussed in [2] [3] does not satisfy this condition because it makes very expensive aggregations over the raw trajectory data in both the temporal and spatial domains. Moreover the lower bound distance function should be defined in such way that allows the application of the incremental approach. This means that it should be possible to reuse the results from one iteration to another.

4.1   Trajectory Approximation and Indexing

To produce the trajectory approximation with the required properties we decided to use of a uniform spatial grid to discretize the spatial domain. Each object location l_i in the stream can be approximated with the grid cell in the boundaries of which it is. Example of a one dimensional trajectory is shown on figure 2. The trajectory shown can be approximated as a sequence of grid cell numbers 1112222333. We can write this approximation in more a compact form $< 1, 3 > < 2, 7 > < 3, 10 >$ by compressing the consecutive object location approximations in the same grid cell $\bar{x}_i$. They are replaced by record $< \bar{x}_i, t_j >$ where t_i is the last time instant in the list of consecutive object location approximations for grid cell $\bar{x}_i$. For example the trajectory on figure 2 has three consecutive location/time instances $u_i = \langle o_i, l_i, t_i \rangle$ all of them in the grid cell 1. These are time instances 1, 2 ad 3. Instead of having a sequence 111 in the approximation we put a record with the grid cell number and the last time instance in the sequence $< 1, 3 >$. More formally a trajectory approximation can be defined as:

Trajectory Approximation. Given trajectory $T(o_i) = \{< l_1, t_1 >, \ldots < l_n, t_n >\}$ of length n, a trajectory approximation is a sequence $\bar{T}(o_i) = \{< \bar{x}_1, \bar{t}_1 >, \ldots, < \bar{x}_m, \bar{t}_m >\}$ where the spatial values contained inside each time frame $(t_{i-1}; t_i)$ are approximated with the grid cell number $\bar{x}_i$ and $1 \leq m \leq n$

Note that this approximation scheme is different from the scheme proposed for the static version of the join problem [2][3], since it approximates the trajectory data only in the spatial domain. We thus avoid the costly approximation along the temporal axis which can be too expensive for the continuous environment.

We proceed with the description of the indexing structure. We propose the creation of a 2-dimensional index space where both dimensions are temporal ("from" and "to" axes) On the "from" temporal axis we plot the time when the object enters a given cell while the "to" temporal axis depicts the time when the object leaves the grid cell. Between these two timestamps the object does not move outside the boundaries of the grid cell. For every record $< \bar{x}_i, \bar{t}_i >$ in the trajectory approximation we place a two dimensional point I_i in the indexing space. We refer to these points as indexing points. A more formal definition of an index point is:

Index point. is a tuple $I_i = \langle o_i, \bar{g}, t_f, t_t, p \rangle$, where o_i is the moving object, $\bar{g}$ is a grid cell number such that $\cap(\bar{g}, T_{o_i}) = true$ for $\forall t \in (t_f, \ldots t_t)$ and p is a pointer to pages on the disk containing the raw trajectory data for time period $(t_f; t_t)$.

For illustration consider the trajectory shown in figure 2. The object stays inside grid cell 1 between time instances 0 and t_1 and has 3 location/time instances $u_i = \langle o_i, l_i, t_i \rangle$ inside this grid cell then it moves to grid cell 2 and stays there between time instances t_1 and t_2. Finally it moves inside grid cell 3. Figure 3 depicts the corresponding indexing space for this example. The object movement is approximated with two *index* points in the indexing space. Both of them show the time period for which the object was inside a given grid cell. For example index point 1 in this 2-dimensional space shows that the

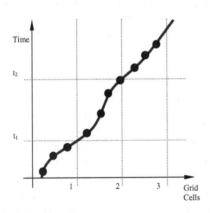
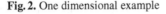
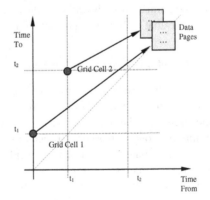

Fig. 2. One dimensional example   **Fig. 3.** Indexing space

moving object was in grid cell 1 in the time interval $(0; t_1)$. Respectively object point 2 shows that moving object was in grid cell 2 in the time interval $(t_1; t_2)$ and so on. A trivial observation is the fact that all index points I_i will be placed above the dashed line on figure 3 which bisects the angle between the two temporal axes (that is because the timestamp when an object leaves an grid cell is bigger than the timestamp when the object enters a grid cell e.g. $\forall I_i; I_i.t_f < I_i.t_t$).

With the above approximation, an object trajectory is transformed to a set of index points in the 2 dimensional indexing space. Inside each index point we keep a pointer to the data page on the secondary level storage which stores the raw trajectory data approximated with this indexing point I_i. For example for index point 1 we keep a pointer to the data page on the disk which has the raw data for time instances 1, 2 and 3. These pointers are used in the verification step when we have to check if the objects indeed satisfy the join criteria using the raw data. Instead of accessing all records for the given trajectory we access only those data pages which have the data for the period of interest. To make the access to the indexing space more efficient we can now use variety of tree-like spatial indexes (R tree or kdb tree) build over all index points in the indexing space.

There are two major advantages of the proposed indexing structure over simple solutions like keeping a trajectory tail for the last δt time instances. First the size of this index is expected to be smaller than the size of an in memory data structure, which holds the fresh trajectory tail. This is because in the proposed index we keep information only for the moments when an object changes its location grid cell instead of keeping all location/timestamp pairs for a period δt. The second advantage is that by issuing a range query inside the index space we can efficiently locate all moving objects which change their location grid cell for the specified time period without accessing all trajectory tails (This will be discussed in detail in section 5).

As time passes, more and more index points will be added to the indexing space. The tree structure built over this space will grow and it performance will eventually deteriorate. Moreover we would like to keep the indexing space and the tree structure over it small enough to fit in main memory for fast access. Possible solution to this problem is to delete all index points I_i from the index space which are too "old". A data older than the time period δt cannot participate as a result so we can safely prune these regions of the indexing space. For example if the time period is $\delta t = 2$ hours then there is no need to keep data older than 2 hours in the structure. The index points, which we can safely remove from the index structure, will be in the shaded region, shown on figure 4.

4.2   Lower Bound Distance Function

Having defined the trajectory approximation we need an appropriate lower bound distance function between the approximated trajectories. Given that the minimal distance between two grid cells is a lower bound of the actual distance between the object locations ($d(\bar{x}_i, \bar{x}_j) \leq d(l_i, l_j)$), as it is shown on figure 5, we can define a lower bound

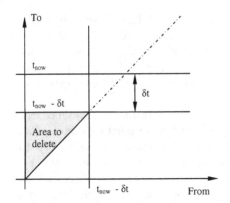

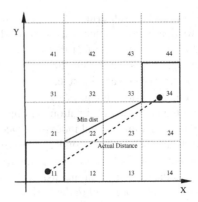

Fig. 4. Indexing space to remove

Fig. 5. Min. distance between two grid cells.

function between the trajectory approximations using the minimal Euclidean distance between the grid cells, i.e.:

$$\bar{\mathcal{D}}_{\delta t, t_i}(\bar{T}(o_{ri}), \bar{T}(o_{si})) = \sqrt{\sum_{i \in (t_i; t_i - \delta t)} d(\bar{r}_i, \bar{s}_i)^2}$$

5   Query Evaluation

We now proceed with the CSTJ evaluation algorithm which assumes a spatiotemporal stream of moving objects approximated in an index structure built as described in section 4.1. There are two major processes in a continuous query evaluation framework that are working in parallel. The first process is keeping the indexed space consistent with the spatiotemporal stream. The second process is responsible for the continuous reevaluation of the CSTJ queries in the system. During its lifetime, a CSTJ query goes through two phases, namely:

– Phase 1. Initial formation of the query result.
– Phase 2. Continuous query reevaluation.

During the first phase the CSTJ query is introduced into the system and the initial result is computed from scratch. Once the initial result of the query is formed, the evaluation of the continuous query moves to the second phase where the query stays in till it is taken out of the system. In this phase the query is reevaluated regularly and the results from the reevaluations are constantly send to the end users. In the remaining of this section we look at each phase in detail.

5.1   Initial Formation of the Query Result

When the query is first introduced into the system, the result has to be computed from scratch. For this phase we modify the "multiple-origin" static join algorithm discussed in [3] to be used with the described indexing scheme. In particular we need to find all pairs of trajectories in the time period $(t_n - \delta t; t_n)$ where the corresponding grid cells for every time instance are not further apart than the threshold ϵ, i.e. $d(\bar{r}_i, \bar{s}_i) \leq \epsilon$, for $i \in (t_n - \delta t; t_n)$.

Each trajectory approximation of length δt can be viewed as a δt-dimensional point in a transformed δt-dimensional space. Using distance function $\bar{\mathcal{D}}_{\delta t}$ defined over the trajectory approximations we can define an ordering of the points in δt-dimensional space by sorting them according to their distances from some set of origins $\bar{O}_i$.

An origin $\bar{O}_i$ is an approximated trajectory with length δt and can be selected arbitrarily. We assign to every trajectory approximation $\bar{T}(o_i)$ a set of q scores $(w_1, \ldots, w_j, \ldots w_q)$, where each score w_j is simply the distance $\bar{\mathcal{D}}_{\delta t}(\bar{O}_j, \bar{T}(o_i))$, between the trajectory approximation $\bar{T}(o_i)$ and origin $\bar{O}_j$. Approximations with different distances from a given single origin $\bar{O}_j$ are considered to be dissimilar. The reverse however is not true: we can have approximations with the same distance from origin $\bar{O}_j$ which are still not spatially close. To reduce the probability of this happening we thus use multiple origins. The verification step however is still needed.

To compute the object scores $(w_1, \ldots, w_j, \ldots w_q)$ we use the index space maintained over the spatiotemporal stream. We locate the portions of the trajectory approximations which belong to the time interval $(t_n - \delta t; t_n)$ by issuing spatial queries in this index space. Given a grid cell and a time interval $(t_f; t_t)$ we can partition the space in four regions, as shown in figure 6.

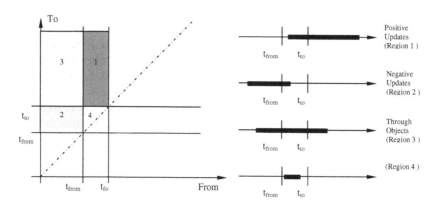

Fig. 6. Index space partitioning

Region 2 contains index points for all these objects which were inside the given grid cell before the beginning of the interval t_f and which moved to another grid cell at some point during the interval $(t_f; t_t)$. Region 1 contains the objects which moved inside the given grid cell at some point during the interval $(t_f; t_t)$ and stayed there until the end of the time interval. Region 3 contains objects which were inside the given grid cell all the time during the interval $(t_f; t_t)$ and region 4 contains objects which moved in and then moved out of the grid cell during the time interval. In regions 1, 2 and 3 for any time period we can have at most one index point I_i per object (for example, having two indexing points in region 3 would mean that the object was at the same time in two different grid cells during the specified period which is a contradiction).

Using spatial queries to locate the index points in these partitions of the index space for time interval $(t_n - \delta t, t_n)$ we can compute the trajectory scores by first computing the squared sum of distances between trajectory approximations and origin $\bar{O}_j$: Let I_1 be the set of indexing points for object o_i in region 1, I_2 the set of indexing points in region 2 and so on. The squared sum of distances between trajectory approximation $\bar{T}(o_i)$ and origin $\bar{O}_j$ will be:

$$\sigma_{t_i, \bar{O}_j}(o_i) = \sum \begin{cases} (I.t_t - (t_i - \delta t))d(I.\bar{g}, \bar{O}_j)^2 \text{ for } I \in I_1; \\ (t_i - I.t_f)d(I.\bar{g}, \bar{O}_j)^2 \text{ for } I \in I_2; \\ (\delta t)d(I.\bar{g}, \bar{O}_j)^2 \text{ for } I \in I_3; \\ (I.t_t - I.t_f)d(I.\bar{g}, \bar{O}_j)^2 \text{ for } I \in I_4. \end{cases}$$

$$w_j(o_i) = \bar{\mathcal{D}}_{\delta t, t_n}(\bar{T}(o_i), \bar{O}_j) = \sqrt{\sigma_{t_n, \bar{O}_j}(o_i)}$$

Since the lower bound distance $\bar{\mathcal{D}}$ is a metric, if two trajectory approximations have at least one w_j score larger than $\sqrt{\epsilon^2 \delta t}$, then there exists at least one time instance t_i in which the corresponding grid cells from the approximations are farther apart than the threshold ϵ. The distance between the grid cells is a lower bound between the actual position of the objects so the corresponding objects do not satisfy the join criteria.

To locate candidate join pairs we sort the trajectory approximations $\bar{T}(o_i)$ using the q scores, in order of $w_1, w_2, ...$, etc. That is, if there is a group of trajectory subsequences having the same value of w_1, they are further sorted on their w_2 score and so on.

Having the trajectory approximation scores $w_1, w_2, ...$, computed and the approximations sorted, we can locate the pairs for which could possibly join. This is done performed by a sliding window algorithm which passes over the sorted list of approximations. We set the size of the window to $2\sqrt{\epsilon^2 \delta t}$ and place the midpoint of the window on the first approximation from dataset s, say $\bar{T}(o_{sj})$, in the sorted list. For all approximations $\bar{T}(o_{ri})$ of dataset r falling inside the window, we compare their scores $w_1, ..., w_q$ with the corresponding scores of $\bar{T}(o_{sj})$. If all of them are within the threshold $\sqrt{\epsilon^2 \delta t}$ we save the pair $\langle o_{sj}, o_{ri} \rangle$ as possible join candidate. Then, we slide the window and place its midpoint on the next element in dataset s and and so on. At the end we verify the generated candidate pairs loading the raw data from the secondary storage. To reduce the number of I/O we follow the pointers inside the index points to locate the data pages storing information for this time period instead of having a full scan over the the pages storing raw trajectory data.

We illustrate the initial formation of the result with an example. Assume that we have 3 moving objects between time instances 1 and 11. Each object report its location every time instance (see figure 7 - the locations where an object reports its position are marked with a dot). We discretize the space with a grid 3x3 where each grid cell is a square with side 10. The minimal distance between the grid cells is given in table 6.1.

The indexing space for this example is shown in figure 8. Next to each index point we have the moving object number o_i to which it belongs and the grid cell number $\bar{x}_j$ in format $\langle o_i, \bar{x}_j \rangle$. Consider a CSTJ query with spatial threshold $\epsilon = 3$ and time period $\delta t = 3$, that is introduced in the system on time instance 6. Object o_1 and object o_3 belong to the first set s and object o_2 belongs to the second set r. The partitioning for time interval $(3; 6)$ is shown in figure 8.

Algorithm 1. CSTJ - Initialization phase

Input: Query $\mathcal{Q} = \{o_r, o_s, \delta t, \epsilon\}$ current time instance t_i
Output: Set of pairs (o_{ri}, o_{sj}) where $T(o_{ri})$ and $T(o_{sj})$ are joined for the last δt time instances
1: Set $\sigma \leftarrow \emptyset, W \leftarrow \emptyset, V \leftarrow \emptyset, Res \leftarrow \emptyset$
2: Find Origins($O_1 .. O_m$);
3: ComputeApproximations($\tilde{O}_1...\tilde{O}_m$);
4: CreatePartitions($t_i - \delta t, t_i$);
5: GetIntexPoints(I_1, I_2, I_3, I_4);
6: **for** each origin O_j in $O_1 .. O_m$ **do**
7:     **for** each moving object o_r in $o_r \cup o_s$ **do**
8:         Compute $\sigma_{t_i, \bar{O}_j}(I_i.o)$
9:         σ.push($\sigma_{t_i, \bar{O}_j}(I_i.o)$);
10: **for** all $\sigma_{t_i, \bar{O}_j}(o_i)$ in σ **do**
11:     $W_{i,j} = \sqrt{\sigma_{t_i, \bar{O}_j}(o_i)}$
12: W.sort()
13: **for** (i = 1; i¡= W.size; i++) **do**
14:     Entry $o_x = W[i]$.objectID
15:     **if** $x \in o_r$ **then**, FindPairsInWindow(o_x, i, W, V, ϵ)
16: **while** V not empty **do**
17:     Entry $< o_{ri}, o_{sj} >= V$.top
18:     **if** $o_{ri} \in o_r$ and $o_{sj} \in o_s$ satisfy the criteria **then**
19:         R.push(o_{ri}, o_{sj})
20: **Return** R

Grid No	11	12	13	21	22	23	31	32	33
11	0	0	10	0	0	10	10	10	14
12	0	0	0	0	0	0	10	10	10
13	10	0	0	10	0	0	14	10	10
21	0	0	10	0	0	10	0	0	10
22	0	0	0	0	0	0	0	0	0
23	10	0	0	10	0	0	10	0	0
31	10	10	14	0	0	10	0	0	10
32	10	10	10	0	0	0	0	0	0
33	14	10	10	10	0	0	10	0	0

We have three indexing points in region 1 (one for each moving object) and three indexing points in region 2. This means that during this time period each object left the grid cell where it was in the beginning of the period and moved into another grid cell. For simplicity we choose a scenario with two origins. Both of them are trajectory approximations with length $\delta t = 3$. Again for simplicity we choose the first one to be $\bar{O}_1 = 33, 33, 33$ and the second one $\bar{O}_2 = 31, 31, 31$. We can then compute the scores for the first trajectory $w_1(o_1) = \sqrt{(4 - (6 - 3))(14)^2 + (6 - 4) * (10)^2} = 19,89$ and $w_2(o_1) = \sqrt{(4 - (6 - 3))(10)^2 + (6 - 4) * (10)^2} = 17,32$. In the same way we compute the scores for the other moving objects $w_1(o_2) = 19,89$, $w_2(o_2) = 17,32$, $w_1(o_3) = 14,14$ and $w_2(o_3) = 0$.

Algorithm 2. FindPairsInWindow

Input: o_x, i, W, V, ϵ

1: $j \leftarrow i$
2: **while** $W[j].scores - o_x.scores < \sqrt{\epsilon^2 \delta t}$ **do**
3:     Entry $o_y = W[j]$
4:     **if** $o_y \in o_s$ **then,** $V.push(o_x, o_y)$
5:     $j --$
6: $j \leftarrow i$
7: **while** $W[j].score - o_x.score < \sqrt{\epsilon^2 \delta t}$ **do**
8:     Entry $o_y = W[j]$
9:     **if** $o_y \in o_s$ **then,** $V.push(o_x, o_y)$
10:     $j ++$

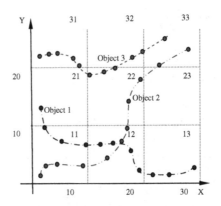

Fig. 7. Moving objects example

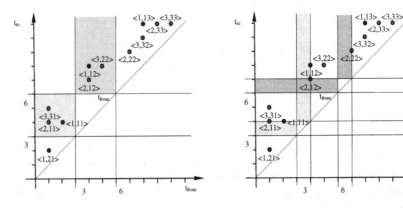

Fig. 8. Partitioning for time interval (3;6)   **Fig. 9.** Time intervals (3;4) and (6;7)

Objects are sorted and placed on a line and then we use the sliding window algorithm where the size of the window is $2\sqrt{\epsilon^2 \delta t} = 2\sqrt{3^2 3} = 5,19$ (figure 10). We place the midpoint of the window on the first s element in the sorted list o_3 we check if there are elements from the second set inside the window. Since there are none we place the window over the next s element o_1. This time there is an element from the set r inside the window - o_2. Thus we report the pair $\langle o_1, o_2 \rangle$ as a candidate pair. There are no more elements in the sorted list so we exit the sliding window algorithm. At the end of the initial evaluation phase we check every candidate pair ($\langle o_1, o_2 \rangle$ in our example) if it indeed satisfies the query criteria.

5.2   Continuous Query Reevaluation

Once the initial result is formed the evaluation of the query moves to its second phase where the query stays active until it is removed from the monitoring system. In this phase we constantly reevaluate the query result and modify it according to the current state of the stream S. To minimize the reevaluation cost we keep the intermediate results produced in every iteration and apply to them the changes which have occurred in the stream. We keep the first $T(o_i).\bar{f}$ and the last grid cell $T(o_i).\bar{l}$ from the trajectory approximations computed in the previous step along with the sum of the squared distances to the origins $\sigma_{t_i, \bar{O}_j}(o_i)$.

Fig. 10. Sliding window algorithm

Assume that the last query reevaluation was at time instance t_p and the current timestamp is t_n. Using the partitioning in the index space shown on figure 3 we can compute the grid cells which form the trajectory approximation for the time period $(t_p - \delta t; t_n - \delta t)$ and those who form the approximation for $(t_p; t_n)$. These portions in the trajectory approximation sustain the difference between the approximations at time instances t_p and t_n. To do so we issue two spatial queries for regions 1, 2 and 4, using time interval $(t_p - \delta t; t_n - \delta t)$ and also time interval $(t_p; t_n)$ to create the partitioning in the index space as it is shown on figure 6. We are focused on these 3 partitions because they contain information about changes in object location during these periods. If there is no change in a location for the period $(t_p - \delta t; t_n - \delta t)$ (e.g there are no indexing points for this object in regions 1,2 and 4 in the partitioning for this time interval) then the object is inside the first grid cell $T(o_i).\bar{f}$ for the whole time interval $(t_p - \delta t; t_n - \delta t)$. The same for time interval $(t_p; t_n)$. If there is no index point in regions 1,2 and 4 for this time period for some object, then the object is still inside grid cell $T(o_i).\bar{l}$. This way keeping the first and the last grid from the grid approximation from the previous time period we avoid the costly spatial query inside region 3 which has the biggest size of all regions.

Using indexing points I_n from the first spatial query along with the last grid cell $T(o_i).\bar{l}$ we compute the sum of squared distances for the interval $(t_p - \delta t; t_n - \delta t)$

$$\Delta_{neg} = \sum \begin{cases} (t_n - t_p)d(T(o_i).\bar{l}, \bar{O}_j)^2 \text{ if } I_{n1} \cup I_{n2} \cup I_{n3} \in \emptyset; \\ (I_1.t_t - (t_p - \delta t))d(T(o_i).\bar{l}, \bar{O}_j)^2 \text{ for } I \in I_{n1}; \\ (t_p - I_1.t_f)d(I_1.\bar{g}, \bar{O}_j)^2 \text{ for } I \in I_{n2}; \\ (I_1.t_t - I_1.t_f)d(I_1.\bar{g}, \bar{O}_j)^2 \text{ for } I \in I_{n4}. \end{cases}$$

This sum of squared distances for $(t_p-\delta t; t_n-\delta t)$ has to be removed from $\sigma_{t_n,\bar{O}_j}(o_i)$ from the previous iteration.

In analogy we compute the sum of squared distances for the period $(t_p; t_n)$ using the indexing points I_p from the second query and the first grid cell $T(o_i).\bar{f}$.

$$\Delta_{pos} = \sum \begin{cases} (t_n - t_p)d(T(o_i).\bar{f}, \bar{O}_j)^2 & \text{if } I_{p1} \cup I_{p2} \cup I_{p3} \in \emptyset; \\ (I_2.t_t - (t_n - \delta t))d(T(o_i).\bar{f}, \bar{O}_j)^2 & \text{for } I \in I_{p1}; \\ (t_n - I_2.t_f)d(I_2.\bar{g}, \bar{O}_j)^2 & \text{for } I \in I_{p2}; \\ (I_2.t_t - I_2.t_f)d(I_2.\bar{g}, \bar{O}_j)^2 & \text{for } I \in I_{p4}. \end{cases}$$

Having $\sigma_{t_p,\bar{O}_j}(o_i)$ from the previous iteration, we can compute the scores for the current time instance

$$\sigma_{t_n,\bar{O}_j}(o_i) = \sigma_{t_p,\bar{O}_j}(o_i) + \Delta_{pos} - \Delta_{neg}$$

$$w_j(o_i) = \bar{\mathcal{D}}_{\delta t,t_n}(\bar{T}(o_i), \bar{O}_j) = \sqrt{\sigma_{t_n,\bar{O}_j}(o_i)}$$

The trajectory scores $w_1, w_2, ...,$ are resorted and processed with the multiple origins sliding window evaluation algorithm to produce the result for time instance t_n. An advantage of this reevaluation schema is that by having a short reevaluation period, the size of regions 1, 2 and 4 in the index partitioning schema will be comparatively small resulting in a limited number of index points accessed during the reevaluation steps.

We will illustrate the reevaluation phase using the same example shown in figure 7. Assume that the query reevaluation is done every time instance and that the current time instance is 7 (e.g one time instance after the initial evaluation). We create the partitioning for time intervals $(3; 4)$ and $(6; 7)$ according to the algorithm. The result is shown on figure 9. There are two indexing points in both regions 2 and 3 (they belong to objects o_1 and o_2) for time interval $(3; 4)$ and one indexing point in the same regions for time interval $(6; 7)$ (generated from object o_2). There are no indexing points for object 3 which means that this object did not change its grid cell during the interval $(3; 4)$ and for this period it was inside grid cell 31 (this is the first grid cell $T(o_3).\bar{f} = 31$ in the trajectory approximation in the previous evaluation step done for time period $(3; 6)$). In analogy object 3 was inside grid cell 22 (which is $T(o_3).\bar{l}$) during the interval $(6; 7)$. We compute $\Delta_{neg}(o_3, \bar{O}_1) = (4 - 3)10^2 = 100$, $\Delta_{neg}(o_3, \bar{O}_2) = (4 - 3)0^2 = 0$, $\Delta_{pos}(o_3, \bar{O}_1) = (7 - 6)0^2 = 0$ and $\Delta_{pos}(o_3, \bar{O}_2) = (7 - 6)0^2 = 0$. So the updated scores for object 3 are $w_1(o_3) = \sqrt{\sigma_{t_6,\bar{O}_1}(o_3) + \Delta_{pos}(o_3, \bar{O}_1) - \Delta_{neg}(o_3, \bar{O}_1)} = \sqrt{200 - 100} = 10$ and $w_2(o_3) = \sqrt{\sigma_{t_6,\bar{O}_2}(o_3) + \Delta_{pos}(o_3, \bar{O}_2) - \Delta_{neg}(o_3, \bar{O}_2)} = \sqrt{0} = 0$. Similarly, $\Delta_{neg}(o_1, \bar{O}_1) = 196$, $\Delta_{neg}(o_1, \bar{O}_2) = 100$, $\Delta_{pos}(o_1, \bar{O}_2) = 100$, $\Delta_{pos}(o_1, \bar{O}_2) = 100$, $\Delta_{neg}(o_2, \bar{O}_1) = 196$, $\Delta_{neg}(o_2, \bar{O}_2) = 100$, $\Delta_{pos}(o_2, \bar{O}_2) = 100$, $\Delta_{pos}(o_2, \bar{O}_2) = 100$ and the new scores for objects o_1 and o_2 $w_1(o_1) = 17.32$, $w_2(o_1) = 17.32$, $w_1(o_2) = 17.32$ and $w_2(o_2) = 17.32$. Then the objects are resorted using the new scores and the sliding window algorithm is re-run.

Algorithm 3. CSTJ - Continuous phase

Input: Query $\mathcal{Q} = \{o_r, o_s, \delta t, \epsilon\}, \sigma, \tilde{O}_1...\tilde{O}_m, t_p, t_n$
Output: Set of pairs (o_{ri}, o_{sj}) where $T(o_{ri})$ and $T(o_{sj})$ are joined for the last δt time instances
1: Set $W \leftarrow \emptyset, V \leftarrow \emptyset, Res \leftarrow \emptyset$
2: CreatePartitions($t_p; t_n$);
3: GetIntexPoints(I_{p1}, I_{p2}, I_{p4});
4: CreatePartitions($t_p - \delta t; t_n - \delta t$);
5: GetIntexPoints(I_{n1}, I_{n2}, I_{n4});
6: **for** each origin O_j in $O_1 .. O_m$ **do**
7:      **for** each moving object o_r in $o_r \cup o_s$ **do**
8:          Compute $\Delta_{pos,i,j}$
9:          Compute $\Delta_{neg,i,j}$
10:          σ.pop($\sigma_{t_p, \bar{O}_j}(I_i.o)$);
11:          $\sigma_{t_n, \bar{O}_j}(I_i.o) = \sigma_{t_p, \bar{O}_j}(I_i.o) + \Delta_{pos,i,j} - \Delta_{neg,i,j}$
12:          σ.push($\sigma_{t_n, \bar{O}_j}(I_i.o)$);
13: **for** all $\sigma_{t_n, \bar{O}_j}(o_i)$ in σ **do**
14:      $W_{i,j} = \sqrt{\sigma_{t_n, \bar{O}_j}(o_i)}$
15: W.sort()
16: **for** (i = 1; i¡= W.size; i++) **do**
17:      Entry $o_x = W$[i].objectID
18:      **if** $x \in o_r$ **then**, FindPairsInWindow(o_x,i,W,V,ϵ)
19: **while** V not empty **do**
20:      Entry $< o_{ri}, o_{sj} >= V$.top
21:      **if** $o_{ri} \in o_r$ and $o_{sj} \in o_s$ satisfy the criteria **then**
22:          R.push(o_{ri}, o_{sj})
23: **Return** R

6   Experimental Evaluation

We proceed with the experimental evaluation of the proposed indexing structure and algorithms for continuous evaluation of CSTJ queries.

6.1   Experimental Environment

In our experiments we use synthetic data to test the behavior of the proposed technique and indexing structure under different settings. We generated synthetic datasets of moving object trajectories. The datasets are generated by simulation using the the freeway network of Indiana and Illinois (see figure 11). We use up to 150,000 objects moving in a 2-dimensional spatial universe which is 1,000 miles long in each direction. The object velocities follow a Gaussian distribution with mean 60 mph, and standard deviation 15 mph. We run simulations for 1000 minutes (time-instants). Objects follow random routes on the network traveling through a number of consecutive intersections and report their position every time-instant. In addition, at least 10% of the objects issue a modification of their movement parameters per time-instant. We choose an R tree with utilization factor 64% as an indexing structure build on the top of the indexing space.

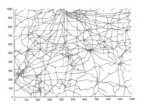

Fig. 11. The map used in the simulations

The average number of time/location tuples $\langle t_i, l_i \rangle$ per index point I is 11 (it is for the selected speed of 60 mph - later in this section we will present experimental results for objects with average speed less than 60 mph). The maximal relative temporal constraint in the queries is set to $\delta t = 40$ minutes. For the maximal time period $\delta t = 40$ minutes, the R tree build over the indexing space has the properties described in the table below.

Property	Value
Tree height	5
Number of nodes	39900
Leaf capacity	20
Index capacity	20

To test the proposed techniques we use two measures: The average number of index node accesses and the average number of data pages per query that need to be retrieved from storage for verification of the result, assuming that one random access is needed for this operation. We also measure the number of trajectory pairs which satisfy the query (e.g the size of the result). We evaluate the query performance in both the initialization and continuous reevaluation phases. Query reevaluation is performed every 2 minutes.

6.2 Experimental Results

Varying the Dataset Size. In the first group of experiments we measure the performance against different data set sizes. We use 4 different datasets with sizes varying from 25,000 up to 150,000 moving objects. The spatial threshold is set to $\epsilon = 30$ miles and the time period δt is set to 20 minutes. The results for the index nodes access, data page access and the number of pairs are shown in figures 12, 13 and 14. As the dataset size increases, the numbers of data pages and index nodes accessed are also growing. As depicted in figure 12, the number of data pages accessed in the initialization and the continuous phases are similar due to the similar number of candidate pairs generated in both phases. The index node access however (figure 13) differs substantially in the two phases. In the continuous phase due to the incremental approach the number of the accessed index nodes is much smaller than in the initial phase. In the continuous phase we do not access the indexing points in region 3 which has the largest size of all 4 regions. Though the distribution of the points in the indexing space is not uniform this region has the biggest number of index points from all four regions in the index space partitioning. As expected, the number of trajectory pairs which satisfy the query is growing with the increase of the dataset size (figure 14).

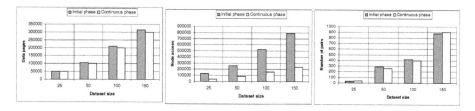

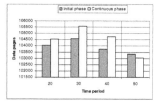 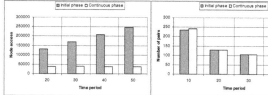

Fig. 12. Size: Data pages   **Fig. 13.** Size: Index nodes   **Fig. 14.** Size: Result set size

Fig. 15. Time: Data pages   **Fig. 16.** Time: Index nodes   **Fig. 17.** Time: Result set size

Varying the Spatial Threshold ϵ. In the next set of experiments we test the behavior of the algorithm for increasing query threshold ϵ (while using a fixed time-interval $\delta t = 20$ minutes). The intuition behind this set of experiments is that by increasing the spatial threshold ϵ the query becomes more relaxed and thus more expensive for evaluation. We use four different values for ϵ varying from 10 to 40 miles. The results are shown in figures 15 16 and 17. As expected with the increase of the threshold we have moderate increase in the number of candidate pairs and the number of result pairs (figure 17). Due to the increased number of reported candidate pairs the data pages accessed (figure 15) also increase since each candidate pair has to be tested using the raw trajectory data. The number of index node accessed however remains constant since the trajectory score computation does not depend on the threshold ϵ (figure 16).

Varying the Time Period δt. In the next group of experiments we tested the behavior of the proposed algorithm for different values of the time period δt varying from 20 up to 50 minutes. We use dataset containing 50,000 moving objects. The spatial threshold is set to $\epsilon = 20$ miles. As it can be depicted from the plots, the number of index nodes accessed during the initialization phase (figure 19) is increasing proportionally to the increase of the time period δt. This proportional increase is due to the fact that larger time periods δt create larger regions 1 and 2 in the space indexing partitioning resulting in larger number of indexing nodes accessed for these two regions. From the plot in figure 18 it can be seen that after time period $\delta t = 30$ minutes the number of raw data I/Os decrease. This is due to the fact that by increasing the length of the query it becomes more restrictive. We have fewer candidates generated and therefore fewer raw data accesses.

The decreased number of candidate pairs results also in a smaller number of pairs in the result set as it can be seen on figure 20.

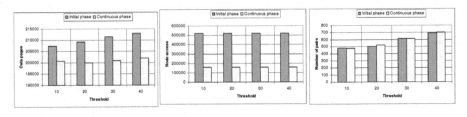

Fig. 18. ϵ: Data pages   **Fig. 19.** ϵ: Index nodes   **Fig. 20.** ϵ: Result set size

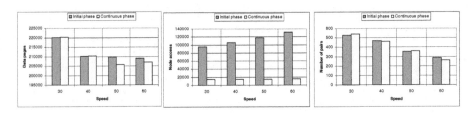

Fig. 21. Speed: Data pages   **Fig. 22.** Speed: Index nodes   **Fig. 23.** Speed: Result set size

Varying the Average Speed of the Moving Objects. All previous experiments were performed using datasets where the average speed is set to 60 mph which is reasonable for a highway traffic. In the last set of experiments we study how the speed of the moving objects affects our algorithm. The intuition here is that by having a slowly moving objects in the system, it will take more time for the object to reach the boundaries of a cell and move to another one. The number of time/location tuples $\langle t_i, l_i \rangle$ per index point I_i is increased and the total number of index points in the indexing space is decreased. We run this set of experiments with four datasets of 100,000 moving objects, where the average speed varies from 30mph to 60mph. The spatial threshold in the query is set to $\epsilon = 30$ miles and the time interval δt is 20 time instances. As expected, the decrease of the average speed in the dataset results in a decrease of the number of indexing nodes accessed (figure 22). The indexing space becomes less dense with the decrease of the average speed. For the same time interval and the same number of moving objects the number of time/location tuples per index point in the 30 mph dataset is 70% from the one in the 60 mph dataset. The number of the trajectory pairs in the result set (figure 23) and the number of data pages I/Os (figure 21) however increase with the decrease of the average speed. This is because in a pair of slow objects, it takes more time for one of them to move on distance ϵ from the second one. So if a pair of slowly moving objects, satisfies the join criteria at one time instance it is more likely to satisfy it in the next time instance. This results in a bigger number of candidate pairs and therefore increased number of raw data I/Os as it can be depicted in figures 21 and 23.

7   Conclusions

We presented an algorithm and an index structure for efficiently evaluating continuous trajectory join queries. Our technique uses compact trajectory representations to build a very small index structure which evaluates approximate answers utilizing a specialized lower bounding distance function. Then, a post filtering step uses only a small fraction of the actual trajectory data before the correct query results can be produced. As future work we plan to extend our techniques for more complex streaming queries with temporal constraints.

References

1. Arge, L., Procopiuc, O., Ramaswamy, S., Suel, T., Vitter, J.S.: Scalable sweeping-based spatial join. In: Proc. of Very Large Data Bases (VLDB), pp. 570–581 (1998)
2. Bakalov, P., Hadjieleftheriou, M., Keogh, E., Tsotras, V.J.: Efficient trajectory joins using symbolic representations. In: Proc. of the International Conference on Mobile Data Management (MDM), pp. 86–93 (2005)
3. Bakalov, P., Hadjieleftheriou, M., Tsotras, V.J.: Time relaxed spatiotemporal trajectory joins. In: GIS 2005: Proceedings of the 13th annual ACM international workshop on Geographic information systems, pp. 182–191 (2005)
4. Beckmann, N., Kriegel, H., Schneider, R., Seeger, B.: The R*-tree: An efficient and robust access method for points and rectangles. In: Proc. of ACM Management of Data (SIGMOD), pp. 220–231 (1990)
5. Brinkhoff, T., Kriegel, H.P., Seeger, B.: Efficient processing of spatial joins using r-trees. In: Proc. of ACM Management of Data (SIGMOD), pp. 237–246 (1993)
6. Faloutsos, C., Roseman, S.: Fractals for secondary key retrieval. In: Proc. of ACM Symposium on Principles of Database Systems (PODS), pp. 247–252 (1989)
7. Gedik, B., Liu, L.: MobiEyes: Distributed Processing of Continuously Moving Queries on Moving Objects in a Mobile System. In: Bertino, E., Christodoulakis, S., Plexousakis, D., Christophides, V., Koubarakis, M., Böhm, K., Ferrari, E. (eds.) EDBT 2004. LNCS, vol. 2992, pp. 67–87. Springer, Heidelberg (2004)
8. Gunadhi, H., Segev, A.: Query processing algorithms for temporal intersection joins. In: Proc. of International Conference on Data Engineering (ICDE), pp. 336–344 (1991)
9. Guttman, A.: R-trees: A dynamic index structure for spatial searching. In: Proc. of ACM Management of Data (SIGMOD), pp. 47–57 (1984)
10. Hjaltason, G.R., Samet, H.: Incremental distance join algorithms for spatial databases. In: Proc. of ACM Management of Data (SIGMOD), pp. 237–248 (1998)
11. Jensen, C.S., Lin, D., Ooi, B.C.: Query and update efficient b+-tree based indexing of of moving objects. In: Proc. of Very Large Data Bases (VLDB) (2004)
12. Koudas, N., Ooi, B.C., Tan, K.-L., Zhang, R.: Approximate nn queries on streams with guaranteed error/performance bounds. In: Proc. of Very Large Data Bases (VLDB) (2004)
13. Koudas, N., Sevcik, K.C.: Size separation spatial join. In: Proc. of ACM Management of Data (SIGMOD), pp. 324–335 (1997)
14. Kwon, D., Lee, S., Lee, S.: Indexing the current positions of moving objects using the lazy update r-tree. In: Proc. of the International Conference on Mobile Data Management (MDM), pp. 113–120 (2002)
15. Lee, M.-L., Hsu, W., Jensen, C.S., Teo, K.L.: Supporting frequent updates in R-Trees: A bottom-up approach. In: Proc. of Very Large Data Bases (VLDB) (2003)
16. Lin, B., Su, J.: On bulk loading tpr-tree. In: Proc. of the International Conference on Mobile Data Management (MDM) (2004)
17. Lin, D., Jensen, C.S., Ooi, B.C., Saltenis, S.: Efficient indexing of the historical, present, and future positions of moving objects. In: Proc. of the International Conference on Mobile Data Management (MDM), pp. 59–66 (2005)

18. Lo, M.-L., Ravishankar, C.V.: Spatial joins using seeded trees. In: Proc. of ACM Management of Data (SIGMOD), pp. 209–220 (1994)
19. Lo, M.-L., Ravishankar, C.V.: Spatial hash-joins. In: Proc. of ACM Management of Data (SIGMOD), pp. 247–258 (1996)
20. Mamoulis, N., Papadias, D.: Multiway spatial joins. ACM Transactions on Database Systems (TODS) 26(4), 424–475 (2001)
21. Mamoulis, N., Papadias, D.: Slot index spatial join. IEEE Transactions on Knowledge and Data Engineering (TKDE) 15(1), 211–231 (2003)
22. Mokbel, M.F., Aref, W.G.: Gpac: generic and progressive processing of mobile queries over mobile data. In: Proc. of the International Conference on Mobile Data Management (MDM), pp. 155–163 (2005)
23. Mokbel, M.F., Xiong, X., Aref, W.G.: SINA: Scalable incremental processing of continuous queries in spatiotemporal databases. In: Proc. of ACM Management of Data (SIGMOD) (2004)
24. Mouratidis, K., Papadias, D., Hadjieleftheriou, M.: Conceptual partitioning: an efficient method for continuous nearest neighbor monitoring. In: Proc. of ACM Management of Data (SIGMOD), pp. 634–645 (2005)
25. Papadopoulos, A., Rigaux, P., Scholl, M.: A performance evaluation of spatial join processing strategies, pp. 286–307 (1999)
26. Patel, J.M., Chen, Y., Chakka, V.P.: Stripes: an efficient index for predicted trajectories. In: Proc. of ACM Management of Data (SIGMOD), pp. 635–646 (2004)
27. Patel, J.M., DeWitt, D.J.: Partition based spatial-merge join. In: Proc. of ACM Management of Data (SIGMOD), pp. 259–270 (1996)
28. Prabhakar, S., Xia, Y., Kalashnikov, D., Aref, W., Hambrusch, S.: Query indexing and velocity constrained indexing: Scalable techniques for continuous queries on moving objects. IEEE Trans. Comput. 51(10), 1124–1140 (2002)
29. Saltenis, S., Jensen, C.S.: Indexing of moving objects for location-based services. In: Proc. of International Conference on Data Engineering (ICDE), pp. 463–472 (2002)
30. Saltenis, S., Jensen, C.S., Leutenegger, S.T., Lopez, M.A.: Indexing the positions of continuously moving objects. SIGMOD Record 29(2), 331–342 (2000)
31. Shan, J., Zhang, D., Salzberg, B.: On spatial-range closest-pair query. In: Hadzilacos, T., Manolopoulos, Y., Roddick, J.F., Theodoridis, Y. (eds.) SSTD 2003. LNCS, vol. 2750, pp. 252–269. Springer, Heidelberg (2003)
32. Tao, Y., Papadias, D., Sun, J.: The tpr*-tree: An optimized spatio-temporal access method for predictive queries. In: Proc. of Very Large Data Bases (VLDB), pp. 790–801 (2003)
33. Xiong, X., Mokbel, M., Aref, W.: Sea-cnn: Scalable processing of continuous k-nearest neighbor queries in spatio-temporal databases. In: Proc. of International Conference on Data Engineering (ICDE), pp. 643–654 (2005)
34. Yu, X., Pu, K.Q., Koudas, N.: Monitoring k-nearest neighbor queries over moving objects. In: Proc. of International Conference on Data Engineering (ICDE), pp. 631–642 (2005)
35. Zhang, D., Tsotras, V.J., Gunopulos, D.: Efficient aggregation over objects with extent. In: Proc. of ACM Symposium on Principles of Database Systems (PODS), pp. 121–132 (2002)

Data Analysis and Integration

Data-Centric Visual Sensor
Networks for 3D Sensing

Mert Akdere, Uğur Çetintemel, Daniel Crispell, John Jannotti,
Jie Mao, and Gabriel Taubin

Brown University, Providence RI 02912, USA

Abstract. Visual Sensor Networks (VSNs) represent a qualitative leap
in functionality over existing sensornets. With high data rates and precise
calibration requirements, VSNs present challenges not faced by today's
sensornets. The power and bandwidth required to transmit video data
from hundreds or thousands of cameras to a central location for processing would be enormous.

A network of smart cameras should process video data in real time,
extracting features and three-dimensional geometry from the raw images
of cooperating cameras. These results should be stored and processed in
the network, near their origin. New content-routing techniques can allow
cameras to find common features—critical for calibration, search, and
tracking. We describe a novel query mechanism to mediate access to
this distributed datastore, allowing high-level features to be described as
compositions in space-time of simpler features.

1   Introduction

We propose an architecture for the construction and use of *Visual Sensor Networks* (VSNs). VSNs will handle much richer data than today's simpler data
collection sensornets. Cameras will perform local image processing, and then
cooperate to perform higher-level tasks, such as calibration, view combination,
object detection, and tracking.

Today, these systems are monitored by a small army of security personnel.
Smart event detection based on higher level analysis of the image data can help
alleviate this burden. Combining information from multiple cameras, space-time
trajectories of individuals can be computed, and suspicious behaviors can be
identified. Today's multi-camera systems perform some of this integration, but
do so in a centralized fashion, requiring all cameras to stream video data to a
single server. These systems will not scale easily to the large networks that could
provide more detailed and comprehensive coverage.

Although complete centralization in untenable, information from several cameras must be combined to make inferences about events occurring in the 3D
world, such as detecting and tracking individuals or vehicles. In order to do so,
the location and orientation of each camera must be determined with respect
to a single coordinate system. This is the camera calibration problem. Mobile
cameras require continuous calibration.

S. Nittel, A. Labrinidis, and A. Stefanidis (Eds.): GSN 2006, LNCS 4540, pp. 131–150, 2008.

To meet the needs of future applications, smart cameras must process video data in real-time, produce lower bit-rate summaries, communicate and share data with neighboring cameras, and execute collaborative algorithms in a decentralized fashion.

1.1 Example Application

Consider how a potential VSN could be deployed and used in a busy metropolitan airport. The network might include the hundreds of static cameras already in use at such an airport today, augmented with thousands of additional static cameras to gain greater coverage. Hundreds of mobile cameras attached to airport personnel and equipment may also play a role.

The VSN will provide security personnel with various ways to access the camera network. The simplest is to ask for views of any area, from any direction. Virtual views would be synthesized from overlapping views provided by the camera network's extensive coverage. Operators might choose to follow people or objects that appear suspicious, or to construct a super-resolution view of a traveler's face. Moving beyond direct human control, such a network could be programmed to draw attention to activity in a restricted area, or an activity by unrecognized personnel. Finally, we envision the network detecting high-level activities such as a traveler who has left his baggage unattended.

It is critical that operators have the tools available to assess the threats detected by the network. For example, users should be able to follow a person or object *back in time* or ask high-level questions about the past. How long has this person been in the room? Which other people has he spoken with? Based on its motion when carried, how heavy is his bag?

1.2 Requirements

In order to support applications of the type we envision, smart cameras must capture and process image data in real-time, and cooperate to make that data available to applications in a structured way.

Multi-Camera Calibration and Time Synchronization. Smart cameras must share a common global coordinate system in order to combine information from disparate cameras in 3D. *Multi-camera geometric calibration* is an active research topic—current solutions are complex, involving cumbersome procedures to overcome the unavoidable partial occlusions, and are based on centralized computation, usually requiring factorization of very large matrices [1,2]. The most common approach is based on *structure from motion* algorithms, in which the pose of all cameras and the location of feature points in 3D are simultaneously estimated. VSNs, on the other hand, require a new robust solution based on distributed algorithms. Furthermore, VSNs with dynamic nodes require new, incremental approaches. *Photometric calibration* is also necessary to account for inevitable differences in sensitivity to light and color between any two image sensors. Finally, VSNs must compensate for the lack of precise *time synchronization* at the frame level.

Virtual Views. Virtual views are images created by integrating visual data from several cameras to simulate the view of a virtual camera where none exists. These views are generated by interpolating sample values from the *light field* [3,4], an abstraction that represents the mapping of 3D rays to colors. In a VSN the individual images that constitute the light field samples are best stored in a distributed fashion near the smart cameras where they are observed. An *image-based routing* protocol efficiently routes virtual view requests to the appropriate smart cameras and composes their individual contributions within the network to minimize network traffic.

Virtual views may also be specified in *resolution* or *time*, leading to virtual video streams. By combining results from several overlapping cameras, a virtual camera of greater frame-rate or resolution may be simulated. Generating video streams consisting of frames defined by similar parameters can be less taxing since the image-based routing mechanism may cache recent decisions. Of particular interest is a virtual video stream that follows a chosen target.

Detection and Tracking. In order to track moving objects, the objects must be segmented from their background. This operation requires a continually maintained model of the background. Once foreground objects are segmented out of the background, noise removal and connectivity analysis defines *blobs*. Tracking 2D blobs over time requires a significant amount of computation at the smart camera level, but reporting their trajectories requires very little communication. Tracking in 3D requires establishing correspondences between blobs detected in separate smart cameras, requiring fine-grained calibration and collaborative processing.

Establishing correspondences between large blobs detected in different images usually reduces to feature detection and matching [5,6]. Features are small blobs which are likely to have a similar appearance in a different image. Features might correspond to corners of buildings, or facial features of people. In general, the feature data interchanged between cameras is not large, but the complexity of feature matching is in principle quadratic in the number of cameras.

Compression. Transmitting high resolution images and high frame-rate video requires a significant amount of bandwidth, and as a result, consumes a significant amount of power. However, multiple cameras capturing multiple views of the same scene produce images and video with significant redundancies. Transmitting this redundant information is a wasteful allocation of the most precious resource of battery operated wireless sensor networks. One approach to remove the redundant information is to reconstruct the 3D structure of the world and transmit a 3D video stream, which can later rendered from an arbitrary point of view. Also, surveillance applications do not require high resolution everywhere in the field of view of the cameras, but just around the detected objects. Image and video compression schemes that interact with object detection algorithms have the potential to reduce the bandwidth utilization significantly.

1.3   Challenges and Contributions

The requirements of a VSN go beyond the techniques developed for existing sensornets for two reasons. First, the raw data is extremely bandwidth intensive. Few sensor systems tackle this challenge. Those that do focus on data types that can be compressed in isolation by, for example, Fourier transform. Second, the image data is difficult to aggregate. Existing systems build collection trees in which aggregation reduces the size of acquired data at join points.

In order to aggregate image data, extensive communication must take place first. Nearby cameras must share image features in order to establish correspondences that create a common coordinate system. Even with aggregation, we expect that in-network storage will be critical to reducing bandwidth requirements. With in-network storage comes challenges in routing and distributed query processing. Our contributions lie in a scheme for storage and processing of data in the VSN, and a high-level data access mechanism for operating on that data.

3D Data-Centric Storage, Routing, and Processing. We introduce data-directed localization to dynamically calibrate without specialized hardware. Nodes will dynamically build ever larger Geographic Hash Tables (GHTs) in which the nodes share a common reference frame. GHTs allow for distributed feature matching and feature matching in smaller GHTs is used to bootstrap localization.

We also introduce data-centric processing (DCP), which places processing elements in the network, located where the data they process will be stored in the GHT. These processing elements operate on data as it becomes available, inserting new, higher-level items into the datastore. Further processing elements may continue this process to produce ever more complex observations.

VSNs must support queries that seek image data for a given object from a given direction. To support these queries that do not map easily to a hash-based content routing scheme, we have developed Image Based Routing. IBR uses a *binmesh* to succinctly represents the views of many cameras in a single summary. Query routing follows the binmeshes down the routing tree toward cameras that observe the target object.

Space-Time Database Abstraction. Our work is intended to simplify the development of 3D sensornet applications in two ways. First, we use a *a space-time "cube" abstraction* for declarative access to the data available in the sensornet. This abstraction hides the raw data acquired by the cameras, providing a form of physical data independence. Applications will "pull" data using SQL-style declarative queries posed on top of the cube abstraction. Second, we rely on a predicate language for specifying space-time *feature patterns* for search and tracking of complex objects and events easily.

Three-Dimensional Hardware. We advocate stereo smart cameras—devices that will include two imaging sensors. The second image sensor adds little to the cost or complexity of the design, but enables significant 3D sensing functionality as well as a reduction in bandwidth utilization by leveraging the 3D structure of the data.

2   System Model

In comparison to existing sensornet approaches, our approach concentrates heavily on in-network processing and storage. Existing systems generally use two techniques: push computation all the way to the sensors (*i.e.*, avoid repeating redundant observations), or thin data by aggregation along a collection tree (*i.e.*, by averaging reading). Our techniques are more cooperative, and less hierarchal. The nodes of VSNs must share the data collected in early stages to enable later staged. Calibration information must be shared to enable feature detection and tracking. Features (and their importance) must be shared in order to guide compression. Collection along a tree is insufficient. Our basic system model is illustrated in Figure 2.

1. "Interesting" data is identified using complex query and event specifications over a space-time database (Sect. 5).
2. Queries and event specifications on application-level features (*e.g.*, a moving person) are translated into an execution plan. The constituent feature-oriented query operators are deployed to allow data centric processing.
3. Image features are extracted locally at each camera. They are then propagated in a neighborhood, using a GHT-based model, to perform localization and calibration in a distributed, collaborative fashion. Furthermore, features are composed to form higher-level features using the data centric execution plan.
4. Raw camera data is acquired, compressed, and stored locally (not necessarily on the source camera). The compression is enhanced because of calibration and feature detection as image redundancies and unimportant data are eliminated. Most sensor networks seek only to produce output for external consumption. VSN nodes must share their computation (here, calibration data and features) in order to function more efficiently (here, compress better).
5. Camera data is retrieved either as direct output from queries, or in a raw form, but identified by the queries (e.g., "here is the trajectory of the car you asked about" or "here is the face of the person who left his bag in the atrium"). To meet potential resource constraints, raw data is compressed

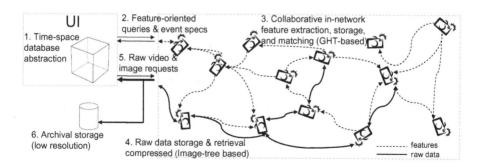

Fig. 1. High level Visual Sensor Network Model

such that interesting features are kept at a higher resolution at the expense of other features.

6. For applications that require it, all data is extracted in compressed, low-resolution format and archived for forensic/historical analysis and legal compliance. Network capacity will limit the fidelity of this data, but calibration and feature knowledge will greatly improve upon the fidelity that might be obtained by individual video streams from each camera.

3   Related Work

Visual Sensor Networks operate at the intersection of many fields, including image processing, traditional wireless networking, and distributed data management.

3.1   Centralized Image Processing

Most existing multi-camera systems are centralized, with all cameras streaming video to a central server where the data is analyzed and visualized [7]. Most early systems focused on the data management benefits resulting from the transition from analog to digital and use the networking infrastructure only as a transport layer [7]. They do no collaborative processing [1,2]. These approaches do not scale to systems with large numbers of video sensors. To address scalability, most of the data intensive processing must be performed at the source, with low-bitrate, highly compressed descriptions of the data transmitted off node. As described above, collaborative processing may result in higher levels of data compression and additional savings in power and bandwidth utilization.

3.2   Routing in Sensornets

A wide variety of protocols have been developed for ad-hoc wireless routing [8,9,10]. These protocols attempt to produce a traditional network-layer, allowing hop-by-hop communication between named end-points. As such, they are not appropriate for the needs of sensornets in which node identity is rarely important.

Instead, specialized protocols have been developed with sensornets in mind. Trickle [11] supports the dissemination of code to a sensornet while minimizing communication costs. Data Centric Storage, in the form of a Geographic Hash Tables [12] (GHTs), has been proposed to allow the storage and retrieval of named data items within the network. A GHT stores a data item by hashing its name to a geographic coordinate and then storing the item (or a pointer) at the node closest to that coordinate. A modified version of GPSR [13] is used to route the item to the nearby node and several replicas. In this work, we extend GHTs in a number of ways adding new support for features that are important to processing data in-network.

3.3   Abstractions for Wireless Sensornets

High-level interfaces for application development in sensor networks have received significant recent attention. Most work in this direction focused on basic operating system and communication support [14], neighborhood and abstract regions [15,16], data-centric event dissemination [17,18], and multi-resolution data storage [19].

Closest to our work are those that take a database-centric approach to sensor network data access, such as TinyDB [20] and Cougar [21]. Recent work [22] has addressed space-time queries in sensornets. These systems focus on aggregation style queries over scalar numerical values, whereas our proposal focuses on a richer space-time database over multi-dimensional image/video data, requiring major changes in the way queries are expressed and executed in the network. In addition, these systems were designed to deal with low-rate data whereas VSNs must handle significantly higher data rates, which we address with novel in-network computation and 3D compression.

3.4   Visual Sensor Networks

The importance of Visual Sensor Networks has been recognized in the broad Sensor Networks literature [23,24], but only relatively small testbed systems, most often wired, have been constructed [25,26]. Similar systems have been proposed for surveillance and security applications, urban traffic control [27], and many more for military applications (refer to [28] for various references). A number of systems for image-based-rendering applications have been proposed including a moderate number of cameras arranged in a regular fashion [29], which can produce super-resolution in time [30]. All of these algorithms are centralized. They require the frames of all cameras to reside in a single place. Attention has not been paid to distributed algorithms applicable to the VSN framework. The same is true for stereo and multi-camera calibration algorithms [1]. Establishing feature correspondences can be done in a pairwise fashion, but it is prone to errors. Robust algorithms such as RANSAC [31] have proven reliable.

4   Network Protocols and Coordination

Visual Sensor Networks have several unusual requirement as compared to traditional wireless networks, or even existing sensornets. Camera networks require fine-grained calibration, distributed feature matching and search, and image oriented routing techniques.

4.1   Data-Directed Localization and Synchronization

We are not the first to observe that sensor networks are data-oriented [32]. In existing sensor networks, requests are routed to the sensors best able to make a specific observation by geographic routing techniques. For example, to find the average temperature in a room, a request is routed toward the room, and then to all nearby sensors. In a static network, this routing might be based on the known locations of immobile sensors. In a mobile network, dynamic routing protocols determine the set of nodes in the target area, usually with the aid of localization hardware such as Cricket [33] or GPS.

In visual sensor networks, the need for localization must be generalized to include orientation and field of view. Small angular errors in orientation may be unacceptable when a distant object's location is estimated, or the views of two cameras are to be integrated.

We propose *data-directed localization* in which smart cameras will localize with respect to each other based on observed image data. Existing localization techniques are not accurate enough to allow for image aggregation. Even the best localizers have only centimeter scale accuracy which is insufficient for accurately determining the orientation of a small sensor node.

In data-directed localization, sensors nodes will detect local features and then cooperate to find common features observed by multiple cameras. When nodes share multiple features, they will orient themselves in a shared coordinate system. Additional cameras will orient themselves in this system by finding features in the shared space. Time synchronization may be accomplished in the same way by the shared observation of temporal events.

Local Feature Detection. Feature detection begins as a local level 2D operation. Next, 3D features can be calculated locally from 2D features by using two image sensors separated by a known baseline. The inter-camera search for correspondences is drastically reduced by using 3D features. Sublinear geometric matching techniques exists for spatial configurations of small groups of features in 3D [34]. These advantages motivate the use of 3D features to reduce bandwidth and power utilization and explain why we advocate the use of smart cameras with two image sensors.

Distributed Feature Matching. Distributed feature matching can be built on top of the idea of a Geographic Hash Table [12]. After detecting local features, each camera using geometric (*not* geographic) hashing [34], to bin them into similar categories. Once categorized, these features are inserted into the GHT, keyed by category. Similar features will therefore be placed at the same location. Prospective matches can be determined at that location, and the nodes with shared features can be notified, allowing them to directly confirm the match and calculate their relative transformation matrix.

Unfortunately, GHTs rely on a preexisting localization scheme to enable geographic forwarding. In order to forward objects to their hashed locations, the nodes must know their own locations and the locations of their neighbors. But we intend to use feature matching in order to determine node locations!

Bootstrapping GHTs. We address this difficulty by bootstrapping localization in small GHTs and extending GHTs to allow merges. Nodes will first organize in small GHTs in which features may be advertised by scoped flooding over n hops. Any two nodes that are within n hops of one another and share features will

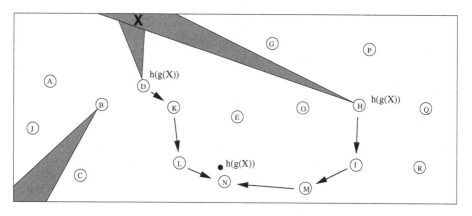

Fig. 2. The feature X is observed by the two distant camera nodes. The feature is categorized through a geometric hash function, g(), and then a storage location is selected with a geographic hash, h(). Each camera routes the feature toward the designated location, where the closest node, N, stores the feature, detects matches, and informs the observers.

detect their overlap and orient themselves in a mutually agreed upon coordinate system. When nodes in separate GHTs detect overlap, the GHTs merge using a single coordinate system. Previous sensornets are incapable of bootstrapping localization in this way because they do not sense distant features in sufficient detail to determine precise relative positions.

In order to merge adjacent GHTs into a single unified coordinate space, shared features must be discovered. However, these features may be shared among nodes that are too far from each other to find each other with scoped broadcast. However, now that small GHTs have been established, they can be used to find more distant matches. A GHT member may be in radio contact with nodes in another GHT. Border nodes from one GHT may place features into the adjacent GHT with knowing the relative transform between the GHTs. Since the GHT will now contain its own features and the features of the adjacent GHT, matches can occur between members of both GHTs.

Feature matching is a useful primitive for tasks beyond localization. To implement tracking, adjacent cameras must realize they are observing the same object. To generate virtual views, multiple views of the same object must be aggregated. Further, we expect most searches in a three dimensional object space to be example based. Such searches can be viewed as feature matching between an abstract target and features detected in the environment. We turn to the task of these general searches in the next section, using feature matching as an important primitive.

We have developed a novel auto-calibration algorithm to estimate the relative position and orientation of several *camera pods*, each consisting of four rigidly mounted network cameras. A processing engine simulated in a computer cluster converted each pod into a multi-sensor smart camera. In addition, basic 3D tracking was demonstrated using the estimated camera parameters. Figure 3 shows

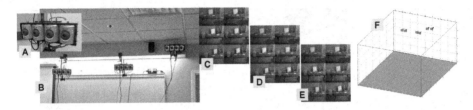

Fig. 3. Preliminary results using proposed 3D feature matching approach: (A) Camera pod; only two of the four cameras used in the calibration algorithm; (B) Experimental setup with three camera pods; (C) features extracted in each image independently; (D) 2D short baseline feature matching within each camera pod results in 3D features; (E) 3D feature matching between camera pods reduces matching complexity; (F) calibration results.

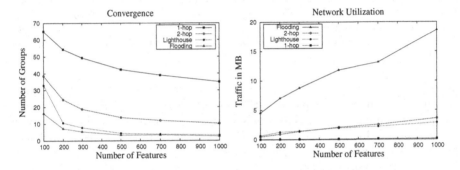

Fig. 4. Simulation results for distributed matching using incrementally built GHTs shows that the GHT scheme exhibits convergence performance comparable to complete feature flooding, but uses little more bandwidth than a simple 2-hop feature propagation protocol. In both graphs, low numbers are better. On the left, they reflect convergence into fewer individual coordinate systems. On the right, the reflect lower network utilization

the pods mounted in the laboratory, and illustrates the steps of the algorithms. Each pod performed small baseline feature matching and stereo reconstruction to construct 3D features. Next, these 3D features were matched between pods. Using a minimum of three correspondences, the pods calculate the rotation and translation necessary to bring themselves into a common reference frame.

In parallel to this exploration of a centralized stereo matching technique, we have simulated the performance of a distributed matching protocol, Lighthouse [35]. Lighthouse attempts to converge an uncalibrated camera network into a single coordinate space using the GHT matching techniques described above. These simulations show that Lighthouse finds distributed matches nearly as well as complete feature exchanges, yet uses less bandwidth than exchanges among 2-hop neighbors. In these simulations of 100 camera networks, complete feature exchange uses approximately five times the bandwidth of Lighthouse. This advantage grows as the network increases in size.

4.2   Feature-Oriented Search and Computation

Visual sensor networks will gather vast amounts of data that must be searched, processed, and acquired by users and applications. GHTs were proposed in traditional sensornets as a compromise between moving all acquired data to a central site, and leaving data at its point of acquisition. Centralizing data requires enormous network capacity and power for transmission, regardless of whether the data is ever queried. Leaving data unindexed at the acquisition site is costly because queries must conduct exhaustive search to locate a data item.

The hash function of a GHT can be thought of as an index on arbitrary data. If data is stored using names that correspond to the needs of queries, retrieval and processing are efficient. For example, if cameras detect and measure the heights of people they observe, they might store these observations (or pointers to them) keyed by those heights, binned in one inch increments. A query can find all individuals greater than six feet tall by examining the hash locations associated with each potential observation above six feet. Sect. 5 presents a relational database abstraction to sensor data, and just as in an RDBMS, we will support arbitrary indexes by storing data according to hash functions that correspond to expected queries.

Data indexed in this way is first hashed into a category, or bin. Next the GHT applies a hash function to select a location for the category. Significant query performance may be lost due to the random locations selected, even for keys that will be queried sequentially. For example, suppose that an application seeks observations of faces in a room—a specific geographic area. These observations might have been inserted into the GHT with keys like, $\langle \texttt{face}, x, y \rangle$, where x and y are the geographic coordinates of the observation, rounded to categorize the observations. The query must lookup each possible value for x and y for coordinates in the room. Hashing each such key results in the storage of these observations arbitrarily throughout the sensornet.

Locality Preserving Hints. We propose widening the interface to the GHT's insertion operation to include an optional coordinate "hint." The GHT hash function will use the supplied coordinate to directly set the high bits of the coordinate at which the data will be stored. In the common case of observations with spatial locality that will be accessed by location, geographic hints will preserve spatial locality and allow queries that access the observations sequentially to operate at a small set of nearby locations. In the example above, observations with similar x and y coordinates will be stored near each other.

Generalizing, we will also allow hints containing a small number of arbitrary scalar values. By taking these values into account during hashing, a set of linear values can be hashed along a line in geographic space. Multiple values can be hashed to a two dimensional patch. Suppose facial observations of faces are stored with hints describing the distance between eyes and the width of mouth. The hinted hash of these observations will map them to a single quadrilateral in the sensornet. Queries that must process a range of values will exhibit spacial locality as they retrieve and process observations in the network.

Feature Aggregation with Data-Centric Processing. The detection of high-level features is generally accomplished by detecting simpler features in a particular arrangement. Primitive feature detectors place a record of their finding in the GHT by inserting the feature under a well-defined name, such as `eye`. To detect higher level features, a second level of feature detectors can be located on the nodes that will receive the individual subfeatures. For example, at hash(`mouth`), a face detector notes the location of the mouth and inserts a partial face observation in the GHT. A similar aggregator creates partial observations for eyes and noses. These observations are all inserted under the well-known name `partial-face` at the same location. When enough observations agree, a face has been detected, and the composite event is inserted at hash(`face`). We consider these operators, placed in the GHT to process values at their insertion point, to be the natural computational analogue to data-centric routing and storage—data-centric processing.

Scoped GHTs. The GHT abstraction provides precise insertion and lookup operations. That is, if any node inserts data under a given key, a lookup from any other node will find the given item. However, feature matching and aggregation do not require this strong guarantee. There is no need for `partial-face` observations associated with eyes observed hundreds of meters apart to be stored at a single point. We propose Scoped GHTs that perform insertions nearby in the case that the observation need not be accessible from afar.

We will explore a geographic hashing scheme, inspired by our previous work on GLS [10], that enable this relaxation. We will divide the world with a fixed sized grid, and store features only in those squares where accessibility is needed. If the feature is used only in queries that require it to be observed within two feet of another feature, then the feature is inserted only in those squares that are within two feet of the feature's location. Scoped insertions require constant power and bandwidth, dependent on square size, rather than the $O(\sqrt{(n)})$ power and bandwidth required to traverse a sensor field of n nodes to a random location.

Figure 5 shows how DCP, using scoped insertions is improved. On the left, observation are sent to a central base station in order to perform complex event detection. On the right, features are are sent only to local base stations. On average, features are tranmitted shorter distances, and the load is spread among many stations. Nonetheless, all complex features are still found, since subfeatures that may be a part of the same larger feature are collected at the same base station.

4.3   Structured Routing and Aggregation

Until now, we have discussed techniques that allow the VSN to perform tasks, such as calibration, feature detection, and tracking, without transmitting large amounts of visual data. Howevever, the VSN must also return visual data efficiently when requested.

Image Based Routing. A sensor network must support data directed queries such as, "Show me the view of the Atrium, from the North." The querier does

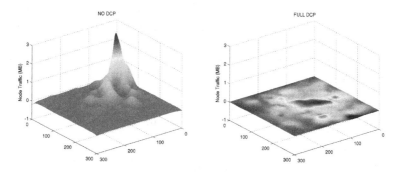

Fig. 5. Network transmissions are plotted as "heat map." On the left, transmission to a centralized base station overwhelm the nearby nodes. On the right, DCP with scoped insertions balances the load to provent hotspots.

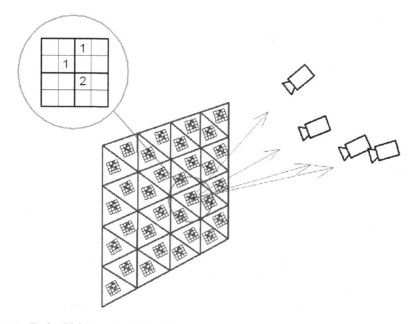

Fig. 6. To build binmeshes, the observable geometry is divided into tiles. A faceted hemisphere is placed on each tile. A *binmesh* describes the set of facets observable by a camera or set of cameras.

not know or care which sensors are involved in answering the query. In a visual sensor network, queries must be routed to the sensors that can observe an area, rather than the sensors in the specific area.

We are integrating our work on Image Based Routing into our VSN framework. The IBR protocol is used when a query is seeking data for a specific location as seen from a specified direction. Image Based Routing resembles a traditional distance-vector routing protocol with route aggregation, rather than hash-based content routing. Image Based Routing is tree based. The leaves of the tree pass descriptions of their fields of view to their parent. As the descriptions work their way toward the root, nodes aggregate multiple view descriptions into a single description that describes their own view and the views of their descendants. A query for any particular view can be routed from the root by choosing the child(ren) that has advertised a view description matching the query. Responses to the query are aggregated from partial responses as they flow back toward the root from the various responding cameras.

View Representation. We have developed the *binmesh*, a datastructure which represents the angles from which a given camera observes a given geometry. Aggregating these binmeshes is, approximately, a bitwise OR. Once aggregated, accuracy remains high. An aggregated binmesh does not represent any new impossible views that its constituent binmeshes did not declare. **Figure 6 is neat.**

5   Data Access and Querying

An important goal of our proposal is to simplify VSN application development. To this end, we will allow users and applications to ask questions about the network in a high-level language that specifies what data to gather from the network without specifying how the query should be executed. The system must adpatively plan in order to execute the query efficiently, taking into account other simultaneous queries, and the fidelity needs that may influence compression.

5.1   Space-Time Database

Our primary abstraction is a space-time 4-dimensional view of the underlying data, consisting of a 3-D volume (x,y,z) representing geographic space and the fourth dimension t representing time. Conceptually, this abstraction captures the data produced by all sensors in a sequence of frozen time frames, where each frame is a 3-dimensional cube that provides a logical model of the world of interest. This abstraction allows users to easily query the system based on spatial attributes on a combination of live and historical data. This view is virtual and not materialized. The implication is that whenever a query is asked on this view, the execution involves accessing the base data stored in a distributed manner in the network. Similar virtual view (but *non* space-time) abstractions have been used by Cougar [21] and TinyDB [20].

Multi-Level Data Representation. Our framework is based on a two-level representation of sensor data: The *raw data layer* and the *view layer*. The raw data layer is the physical layer that continuously acquires and stores camera data. The view layer is the logical layer that transforms raw sensor data into the cube abstraction. User queries are executed on this abstraction. This layering provides physical data independence, a key concept borrowed from traditional relational database systems, which shields users and applications from the details of our data-centric protocols.

User queries can be one time or continuous and can be saved as named views that can be reused once defined. This style of cascading is similar to the way that views are cascaded in traditional database systems. In simplified terms, the semantics of cascading of a query q and a view v is that the output resulting from the execution of v will be fed into q. Cascading simplifies the expression of complicated queries and allows the same underlying query to be used concurrently by multiple others, facilitating resource and result sharing. Furthermore, multiple cascading queries allows for interesting cross-query optimizations (*e.g.*, pushing decimation operations present in the high-level query to the underlying query during execution).

Data Access Methods. The basic data access and querying interface will be a linear, SQL-like notation from declarative queries. Consider the following example query:

```
SELECT from CUBE
 WHERE location = [(50,50,50), (100, 100, 100)]
 VIEWPOINT = (100, 100, 100)
 WITH RESOLUTION 20 fps
 SAVE AS VIEW ''Atrium NE''
```

This continuous query selects a volume of space specified by two corners of a sub-cube and asks for an image stream that corresponds to the target volume as observed from a specific viewpoint. The data are to be acquired with a temporal resolution of 20 frames per second. If the viewpoint cannot be presented due to lack of data, the system might offer an alternative but similar viewpoint for which data is available. The query is saved as view "Atrium NE" as it provides a view of the room named Atrium from the North-East direction. On top of this view, we can define another query that returns images containing bag-like shapes with a resolution of ten frame per second:

```
SELECT  from ''ATRIUM NE''
 WHERE object = ''bag''
 RESOLUTION 10 fps
```

We envision moving beyong a textual language to a graphical tool to allow incremental visualization of the sensornet data. The interface of the tool will resemble the familiar mapping software in that it will allow users to graphically select geographical spaces, zoom in and out, pan in different directions, as well as provide more advanced features such as selecting arbitrary viewpoints, and looking back in time. The operations specified through the visualizer will be translated into queries and submitted for execution to query the sensornet.

Space-Time Feature Predicates. Any interesting VSN will require search and detection of objects and activities (events) based on images. A user might be interested in finding where a specific person was at a specific point in time, locating all bags of a certain color, size, and shape, or even ask to see the people that are currently running. In general, there is a clear need for an extensible programming framework that will facilitate the specification of objects and activities of interest.

Our framework will give the users the ability to specify spatial and temporal features on image data. We uniformly represent both objects and activities using features. Spatial features are defined based on relationships of data over space. For example, a head can be described using a spatial relationship among other (lower-level) features such as eye, nose, ear, mouth, etc. Such relationships are models that represent a feature using the relative spatial orientation of one or more lower-level features (*e.g.*, a nose is below the eyes and above the mouth). Users will register predicates that evaluate both primitive features (*e.g.*, a nose or an eye) and composite features (*e.g.*, head, body, person). Temporal features will be defined in a similar manner although, in this case, one is interested in the position of features over time. For example, the activity of "moving" can be expressed as a specific feature changing its location over time.

Clearly spatial and temporal predicates can be intermixed to express arbitrarily complex objects and activities: a running person can be identified by evaluating the spatial predicate that detects a person in a given time snapshot and then a temporal predicate that checks whether that person is moving faster than a given threshold. Once defined, the predicates will be sent to the network locations where they will be evaluated with data centric processing. Defining objects and events using feature predicates will allow us to use the feature-based routing and matching techniques outlined in Sec. 4.1 as the uniform underlying in-network query execution mechanism.

We have built a space-time database layer that facilitates queries over the location and trajectory of moving objects. Object positions acquired using cameras and Cricket nodes are stored in a centralized database and organized in multiple orders to facilitate efficient space-time queries. This work is explores sophisticated interactions with people and various artifacts in a large museum setting. This Smart Museum project will continue to serve as a target application for our work.

Adaptive Query Execution. Once the user submits a query to the system, the query will be translated into an execution plan. The planning phase must decide, based on the query specification, which routing indexes (spatial, temporal, feature-based, or a combination) and feature detectors to use. The query plan will be sent to the network and executed collectively by the appropriate nodes (as described earlier in this section) in a distributed fashion, after which results are sent back to the user. This is a distributed query optimization problem and is one of the main challenges that we will tackle.

Query execution must also *adapt* to dynamically changing workload and network characteristics. Our primary tool for adaptation will be *application-aware compression* where the novel compression techniques discussed earlier will be applied to camera data selectively based on their utility to the existing queries and the availability (or lack of) resources. The issues we will address include how to extract utility information from the existing workload and how to use this information to guide compression decisions.

5.2   Image and Video Stream Compression

We regard 3D reconstruction as a mechanism to compress the data captured from multiple video streams, resulting in lower power consumption. Polygon Meshes and oriented point clouds are popular representations for 3D surfaces in computer graphics. Polygon meshes are highly compressible [36,37,38,39,40,41,42,43], and point clouds are error resilient. In the context of VSNs a representation with the two properties is needed. We propose a new surface representation composed of time-dependent overlapping parameterized surface patches, or *3D video streams*.

3D video streams as compression of multiple 2D video streams. Traditional video standards such as MPEG-4 support the transmission of a dynamic 3D scene as a set of multiplexed video streams, one per camera. As the number of video streams grows under constant channel capacity, the overall image quality decreases. We advocate the generation of a compressed representation of all possible views, exploiting the correlation between multiple views of the same scene. Desired views are rendered at the terminal. More computation may be required both at the encoder and decoder, but overall distortion will be minimized, and power consumption due to data transfer will be significantly reduced.

Adaptive 3D stream and Video Sampling. Figure 7 illustrates the way our compression may leverage feature knowledge obtained by the VSN. This will extend the work of Balmelli, Taubin, and Bernardini [44] from static textured polygon meshes to 3D video streams. A smart camera captures a high resolution video stream. In cooperation with other cameras, queries detect and track objects, such as faces or suitcases, resulting in a monochromatic *alpha channel* which assigns *importance values* to different pixels. The smooth alpha channel can be downsampled quite aggressively. Each of these decimated frames is used to generate a 2D *warping function* used to resample the frames of the source video stream adaptively on a pixel grid of the same dimensions as the downsampled alpha channel. The result is transmitted as an RGBA video stream of low resolution and normal frame rate, which can be further compressed using standard methods. These low resolution images preserve the details of the regions of interest at the captured resolution. In the decoder, the inverse warping function is computed from the alpha channel, and the unwarped image is recovered at the full

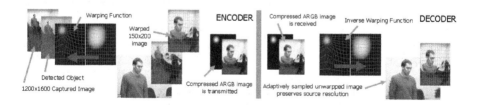

Fig. 7. Video Compression for Surveillance will use adaptive sampling and downsampling

resolution in the important regions. Again, by sharing the results of prior stages (feature detection) greater efficiencies can be obtained later (in compression).

6 Conclusions

VSNs represent an opportunity and challenges. Smart cameras offer far richer capabilities than simpler sensors, but require far greater effort to coordinate effectively. We have outlined a vision for using camera networks effectively, from the initial problem of self-calibration, through feature and image retrieval, to an expressive and efficient query language for application interaction.

References

1. Svoboda, T., Martinec, D., Pajdla, T.: A convenient multi-camera self-calibration for virtual environments. PRESENCE: Teleoperators and Virtual Environments 14(4) (August 2005)
2. Wang, F.Y.: An Efficient Coordinate Frame Calibration Method for 3-D Measurement by Multiple Camera Systems. IEEE Transactions on Systems, Man, and Cybernetics – Part C: Applications and Reviews 35(4), 453–464 (2005)
3. McMillan, L., Bishop, G.: Plenoptic Modeling: An Image Based Rendering System. In: Siggraph 1995. Conference Proceedings, pp. 39–46 (1995)
4. Levoy, M.: Light Field Rendering. In: Siggraph 1996. Conference Proceedings, pp. 31–42 (1996)
5. Mikolajczyk, K., Schmid, C.: A performance evaluation of local descriptors. IEEE Transactions on Pattern Analysis and Machine Intelligence 27(10), 1615–1630 (2005)
6. Lowe, D.: Distinctive image features from scale invariant features. International Journal of Computer Vision 60(2), 91–110 (2004)
7. Valera, M., Velastin, S.: Intelligent distributed surveillance systems: a review. IEE Proceedings on Vision, Image, and Signal Processing 152(2), 192–204 (2005)
8. Perkins, C., Royer, E., Das, S.R.: Ad hoc On demand Distance Vector (AODV) routing. Internet draft (work in progress), Internet Engineering Task Force (October 1999)
9. Johnson, D.B.: Routing in ad hoc networks of mobile hosts. In: Proc. of the IEEE Workshop on Mobile Computing Systems and Applications, pp. 158–163 (December 1994)
10. Li, J., Jannotti, J., Couto, D.S.J.D., Karger, D.R., Morris, R.: A scalable location service for geographic ad hoc routing. In: Proc. ACM/IEEE MobiCom (August 2000)
11. Levis, P., Patel, N., Culler, D., Shenker, S.: Trickle: A self-regulating algorithm for code propagation and maintenance in wireless sensor networks. In: NSDI 2004 (March 2004)
12. Ratnasamy, S., Karp, B., Yin, L., Yu, F., Estrin, D., Govindan, R., Shenker, S.: GHT: A geographic hash table for data-centric storage in sensornets. In: Proc. of the 1st ACM International Workshop on Wireless Sensor Networks and Applications (WSNA) (September 2002)
13. Karp, B., Kung, H.T.: GPSR: greedy perimeter stateless routing for wireless networks. In: Proc. ACM/IEEE MobiCom, pp. 243–254 (August 2000)

14. Levis, P., Madden, S., Gay, D., Polastre, J., Szewczyk, R., Woo, A., Brewer, E., Culler, D.: The emergence of networking abstractions and techniques in TinyOS. In: NSDI 2004 (2004)

15. Welsh, M., Mainland, G.: Programming sensor networks using abstract regions. In: NSDI 2004 (2004)

16. Whitehouse, K., Sharp, C., Brewer, E., Culler, D.: Hood: a neighborhood abstraction for sensor networks. In: MobiSys 2004 (2004)

17. Intanagonwiwat, C., Govindan, R., Estrin, D.: Directed diffusion: A scalable and robust communication paradigm for sensor networks. In: Proceedings of the Sixth Annual International Conference on Mobile Computing and Networking (Mobi-COM 2000), Boston, Massachussetts (2000)

18. Heidemann, J., Silva, F., Intanagonwiwat, C., Govindan, R., Estrin, D., Ganesan, D.: Building efficient wireless sensor networks with low-level naming. In: Proceedings of the Symposium on Operating Systems Principles, Lake Louise, Banff, Canada (2001)

19. Ganesan, D., Greenstein, B., Perelyubskiy, D., Estrin, D., Heidemann, J.: An evaluation of multi-resolution storage for sensor networks. In: Proceedings of the ACM SenSys Conference (2003)

20. Madden, S., Franklin, M.J., Hellerstein, J., Hong, W.: TAG: A tiny aggregation service for ad-hoc sensor networks. In: Proceedings of the 5th USENIX Symposium on Operating Systems Design and Implementation (OSDI 2002), Boston, Massachusetts (December 2002)

21. Yao, Y., Gehrke, J.: Query processing in sensor networks. In: Proc. of the First Biennial Conference on Innovative Data Systems Research (CIDR 2003) (January 2003)

22. Coman, A., Nascimento, M., Sander, J.: A framework for spatio-temporal query processing over wireless sensor networks. In: 2nd International VLDB Workshop on Data Management for Sensor Networks (2005)

23. Obraczka, K., Manduchi, R., Garcia-Luna-Aveces, J.: Managing the information flow in visual sensor networks. In: Proceedings of the 5th. International Symposium on Wireless Personal Multimedia Communications (October 2002)

24. Wolf, W., Ozer, B., Lv, T.: Smart cameras for embedded systems. IEEE Computer 35(9), 48–53 (2002)

25. Trivedi, M., Mikic, I., Bhonsle, S.: Active Camera Networks and Semantic Event Databases for Intelligent Environments. In: Proceedings of the IEEE Workshop on Human Modeling, Analysis and Synthesis, Hilton Head, South Carolina (June 2000)

26. Hampapur, A., Brown, L., Connell, J., Pankanti, S., Senior, A., Y-L, T.: Smart surveillance: Applications, technologies and implications. In: Proceedings of the IEEE Pacific-Rim Conference On Multimedia, Singapore (December 2003)

27. Esteve, M., Palau, C., Catarci, T.: A Flexible Video Streaming System for Urban Traffic Control. IEEE Multimedia 13(1), 78–83 (2006)

28. ACM 2nd International Workshop on Video Surveillance & Sensor Networks

29. Wilburn, B., Smulski, M., Kelin Lee, H.H., Horowitz, M.: The light field video camera. In: SPIE Electronic Imaging 2002, Media Processors, Conference Proceedings (2002)

30. Wilburn, B., Joshi, N., Vaish, V., Levoy, M., Horowitz, M.: High speed videography using a dense camera array. In: Proceedings of the IEEE Conference on Computer Vision and Pattern Recognition (CVPR 2004), pp. 294–301 (2004)

31. Lacey, A.J., Pinitkarn, N., Thacker, N.A.: An Evaluation of the Performance of RANSAC Algorithms for Stereo Camera Calibration. In: Proceedings of The Eleventh British Machine Vision Conference (September 2000)
32. Shenker, S., Ratnasamy, S., Karp, B., Govindan, R., Estrin, D.: Data-centric storage in sensornets. In: Proc. 1st Workshop on Hot Topics in Networking (HotNets-I) (October 2002)
33. Priyantha, N., Chakraborty, A., Balakrishnan, H.: The Cricket location-support system. In: Proc. ACM/IEEE MobiCom (August 2000)
34. Wolfson, H.J., Rigoutsos, I.: Geometric hashing: An overview. In: IEEE Computational Science and Engineering, pp. 10–21 (October-December 1997)
35. Jannotti, J., Mao, J.: Distributed calibration of smart cameras. In: Proc. Workshop on Distributed Smart Cameras (DSC 2006) (2006)
36. Guéziec, A., Taubin, G., Horn, B., Lazarus, F.: A framework for streaming geometry in vrml. IEEE Computer Graphics and Applications, 68–78 (March/April 1999)
37. Guéziec, A., Taubin, G., Lazarus, F., Horn, W.: Converting sets of polygons to manifold surfaces by cutting and stitching. In: IEEE Visualization 1998 Conference Proceedings, pp. 383–390 (October 1998)
38. Guéziec, A., Bossen, F., Taubin, G., Silva, C.: Efficient compression of non-manifold meshes. In: IEEE Visualization 1999 Conference Proceedings (October 1999)
39. Guéziec, A., Taubin, G., Lazarus, F., Horn, W.: Simplicial maps for progressive transmission of polygonal surfaces. In: VRML 1998. ACM Press, New York (1998)
40. Taubin, G.: A signal processing approach to fair surface design. In: Siggraph 1995 Conference Proceedings, pp. 351–358 (August 1995)
41. Taubin, G., Horn, W., Lazarus, F., Rossignac, J.: Geometric Coding and VRML. Proceedings of the IEEE 86(6), 1228–1243 (1998)
42. Taubin, G., Guéziec, A., Horn, W., Lazarus, F.: Progressive forest split compression. In: Siggraph 1998 Conference Proceedings, pp. 123–132 (July 1998)
43. Taubin, G., Rossignac, J.: Geometry compression through topological surgery. ACM Transactions on Graphics 17(2), 84–115 (1998)
44. Balmelli, L., Taubin, G., Bernardini, F.: Space-Optimized Texture Maps. Computer Graphics Forum 21(3) (September 2002)

A Vision for Cyberinfrastructure for Coastal Forecasting and Change Analysis

G. Agrawal, H. Ferhatosmanoglu, X. Niu,
K. Bedford, and R. Li

Ohio State University, Columbus OH 43210

Abstract. This paper gives an overview of a recently initiated cyberinfrastructure project at The Ohio State University. This project proposes to develop and evaluate a cyberinfrastructure component for environmental applications. This will include developments in middleware, model integration, analysis, and mining techniques, and the use of a service model for supporting two closely related applications. These applications will be real-time coastal nowcasting and forecasting, and long-term coastal erosion analysis and prediction.

1 Introduction

Over the years, much work has been done on observing and modeling the environment. Many complex systems have been, or are being, built. An example of such an effort is the Integrated Ocean Observing System (IOOS)[1], which is being built with a number of goals, including detecting and forecasting oceanic components of climate variability, ensuring national security, among others.

Despite advances in the amount of data being collected (including larger number of sources as well as increased spatio-temporal granularity) and enhancements in the techniques being developed for analyzing these datasets, we believe that a number of challenges remain in this area:

- The current systems are very tightly coupled. There is hardly any reuse of algorithm implementations across different systems. It is also extremely hard to test or incorporate new analysis algorithms.
- The implementations are closely tied to the available resources.
- The existing systems cannot adapt the granularity of analysis to the resource availability and time constraints.

The emerging trend towards (closely related) concepts of service-oriented architectures [9] and grid computing [12] can alleviate the above problems. They can enable development of services which are not tied to specific datasets or end applications, and implementation of applications using these services. However, this also requires advances in grid middleware components that can support streaming applications and data virtualization/integration.

---

[1] http://www.ocean.us

S. Nittel, A. Labrinidis, and A. Stefanidis (Eds.): GSN 2006, LNCS 4540, pp. 151–174, 2008.

This paper gives an overview of a recently initiated project at The Ohio State University. This project proposes to develop and evaluate a cyberinfrastructure component for environmental applications. This will include developments in middleware, model integration, analysis, and mining techniques, and the use of a service model for supporting two closely related applications. These applications will be real-time coastal nowcasting and forecasting, and long-term coastal erosion analysis and prediction.

1.1 Overview of the System Being Built

We view the cyberinfrastructure software support for environmental applications as comprising four layers:

- At the lowest level, we have basic grid middleware services: Globus (which provides resource monitoring and security) and related middleware standards and services, including Grid Data Access and Integration (DAI) standards. This work has been developed and supported by programs like the NSF Middleware Initiative (NMI).
- At the second level, we have three advanced data-intensive middleware services developed at Ohio State. The particular components will be:
  - GATES (Grid-based Adaptive Execution on Streams) [4]: This middleware allows development of grid-based streaming applications, which can adapt the processing granularity to meet real-time constraints. Continuous sensor-based data is available for most environmental applications. There are many situations where one needs to react on a real-time basis, for example, when there is an oil spill in a lake.
  - Data Virtualization and Wrapper Generation Middleware [39]: The goal here is to make applications or application components independent of the specific data formats. Our work on data virtualization allows application to query or process a complex dataset with a simpler or abstract view, e.g., a relational table based view. Our work on wrapper generation allows data and tools with different formats to be integrated.
  - FREERIDE-G (Framework for Rapid Implementation of Datamining Engines in a Grid): This system allows parallel implementation of data mining or data-intensive scientific applications that involve data in a remote repository.
- At the third level, specific algorithms and data analysis techniques will be implemented as *grid services*, i.e., they will be implemented so that they can be accessed by different application developers, can be applied on different data sources, and also, can be executed on a variety of resources. Using the tools from the previous layer will help in achieving these goals. The specific services in our implementation will be:
  - Multi-model/Multi-sensor 3D mapping, where novel algorithms will be coupled with our wrapper generation service to fuse data collected at different times and altitudes, and with different imaging principles.

- Query planning service, where the focus will be to extract and use appropriate metadata from image datasets.
- Spatio-temporal mining services, which include algorithms for both offline, scalable analysis (implemented using FREERIDE-G) and distributed streaming analysis (implemented using GATES).

— Finally, at the top-most level, we have the end applications. These will be developed using the services from previous layers. The two applications we will target are real-time coastal forecasting/nowcasting, and long term coastal analysis and prediction.

Overall, this project will achieve the following goals. It will demonstrate an architecture for constructing cyberinfrastructure for environmental applications. It will also contribute to further development of our three middleware systems, and will apply and evaluate them for challenging real applications. New algorithms will be developed for multi-sensor/multi-model data fusion, for extracting metadata for image applications and using it for distributed query planning, and for data mining on spatio-temporal heterogeneous datasets. Moreover, these algorithms will be made available as grid services. This project will also enable developments in the area of coastal informatics, particularly, in terms of enabling flexibility and adaptivity in resource utilization, and in using advanced models and analysis algorithms.

2   Advanced Middleware Systems

The characteristics of the applications in the environmental domain, as well as in many related areas, pose many challenges for grid middleware systems. Two of the most common problems are related to supporting real-time or near real-time response to distributed data streams, and integrating data from a large number of sources.

Ongoing work at The Ohio State University has been developing grid middleware components which address these problems. In this section, we describe three systems, each of which will provide a key functionality in our proposed cyberinfrastructure component for environmental applications. These systems are developed on top of existing basic middleware services, i.e. Globus and related standards, and will enable development of more specialized services for environmental applications, including services for multi-sensor 3D mapping of data, distributed query processing, and mining.

The first two of these systems provide support for adaptive and resource-aware execution on distributed streaming data, and data integration and virtualization, respectively. The third system will enable development of scalable data mining and data processing applications in grid environments.

2.1   GATES: A Grid-Based Middleware for Streaming Data

This section describes the motivation and the major design aspects of the GATES (Grid-based AdapTive Execution on Streams) system that has been developed at The Ohio State University [4].

Motivation. Increasingly, a number of applications across a variety of science and engineering disciplines rely on, or can potentially benefit from, analysis and monitoring of *data streams*. In the stream model of processing, data arrives continuously and needs to be processed in *real-time*, i.e., the processing rate must match the arrival rate. Increasing numbers of high precision data collection instruments and sensors that are generating data continuously, and at a high rate, have contributed to this model of processing. The important characteristics that apply across a number of stream-based applications are: 1) the data arrives continuously, 24 hours a day and 7 days a week, 2) the volume of data is enormous, typically tens or hundreds of gigabytes a day, and the desired analysis could also require large computations, 3) often, this data arrives at a distributed set of locations, and all data cannot be communicated to a single site, and 4) it is often not feasible to store all data for processing at a later time, thereby, requiring analysis in *real-time*; alternatively, timely response requires that the data be processed in real-time or *near real-time*. An environmental application like coastal forecasting and nowcasting clearly involves streaming data and the need for real-time response. However, in the past, grid technologies have not been used for such applications.

GATES system is based on need for supporting flexible and adaptive processing of distributed data streams using grid technologies and standards.

Key Goals. There are three main goals behind the design of the GATES system.

1. Enable the application to achieve the best accuracy, while maintaining the *real-time* constraint. For this, the middleware allows the application developers to expose one or more *adaptation* parameters at each stage. An *adaptation* parameter is a tunable parameter whose value can be modified to increase the processing rate, and in most cases, reduce the accuracy of the processing. Examples of such adaptation parameters are, rate of sampling, i.e., what fraction of data-items are actually processed, and size of summary structure at an intermediate stage, which means how much information is retained after a processing stage. The middleware automatically adjusts the values of these parameters to meet the real-time constraint on processing. This is achieved through a *self-adaptation* algorithm.

2. Support distributed processing of one or more data streams, by facilitating applications that comprise a set of *stages*. For analyzing more than one data stream, at least two stages are required. Each stage accepts data from one or more input streams and outputs zero or more streams. The first stage is applied near sources of individual streams, and the second stage is used for computing the final results. However, based upon the number and types of streams and the available resources, more than two steps could also be required. All intermediate stages take one or more intermediate streams as input and produce one or more output streams. GATES's APIs are designed to facilitate specification of such stages.

3. Enable easy *deployment* of the application. This is done by supporting a *Launcher* and a *Deployer*. The system is responsible for initiating the different stages of the computation at different resources. The system also allows

the use of existing grid infrastructure. Particularly, the current implementation is built on top of the Open Grid Services Infrastructure (OGSI) [11], and uses its reference implementation, Globus Toolkit (GT) 3.0.

GATES is also designed to execute applications on heterogeneous resources. The only requirements for executing an application are: 1) support for a Java Virtual Machine (JVM), as the applications are written in Java, 2) availability of GT 3.0, and 3) a web server that supports the user application repository. Thus, the applications are independent of processors and operating systems on which they are executed. Further details of how our middleware uses GT 3.0 are documented in an earlier publication [4]. In the future, we expect to further integrate GATES with the WS-Resource Framework and Globus 4.0.

System Architecture and Design. The overall system architecture is shown in Figure 1. The system distinguishes between an *application developer* and an *application user*. An application developer is responsible for dividing an application into stages, choosing adjustment parameters, and implementing the processing at each stage. Moreover, the developer writes an XML file, specifying the initial configuration information of an application. Such information includes the number of stages, locations of data sources and destination, and where the stages' codes are. After submitting the codes to application repositories, the application developer informs an application user of the URL link to the initial configuration file. An application user is only responsible for starting and stopping an application.

The above design simplifies the task of application developers and users, as they are not responsible for allocating grid resources and initiating the different stages at different resources. To support convenient deployment and execution, the *Launcher*, the *Deployer* and the *Grid Resource Manager* are used. The procedure for launching an application is as follows. To start an application, the

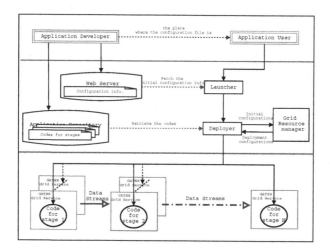

Fig. 1. Overall System Architecture for GATES

user simply passes the XML file's URL link to the Launcher. The Launcher is in charge of fetching the XML file and passing it to the Deployer. The Deployer is responsible for the deployment. Specifically, it 1) analyzes the initial configuration information sent by the Launcher, 2) consults with the Grid Resource Manager to get a complete *deployment configuration* (defined in the next section), 3) initiates instances of GATES *grid services* according to the deployment configuration, 4) retrieves the stage codes from the application repositories, and 5) uploads the stage specific codes to every instance, thereby customizing it.

After the Deployer completes the deployment, the instances of the GATES grid services start to make network connections with each other and execute the stage functionalities. The GATES grid service is an OGSA Grid service [10] that implements the self-adaptation algorithm and is able to contain and execute user-specified codes.

2.2   Bridging Data Format Differences: Data Virtualization and Automatic Wrapper Generation Approach

One of the major challenges in developing applications that process data from a large number of sensors (and other sources) is that this data exists in a variety of low-level formats. This has (at least) two consequences. First, even when processing data from a single source that produces data in a complex and low-level format, an application developer needs to put in a significant effort in understanding the layout of data. The specification of the processing also becomes more involved, and closely dependent on the format of the data. Second, integration and processing of data from a large number of sources becomes even harder. Advances in sensor technologies also result in new data sources, and/or a change in format of the data being produced by an existing source. Current distributed data processing applications are not very well suited for adjusting to such changes.

Data integration is a well-studied problem in computer science, and a number of approaches have been developed. One common approach is to use *wrapper* programs, which can transform the data from one source into a format that is understandable by another. OPeNDAP [5], a system widely used in environmental and oceanographic applications, provides data virtualization through a data access protocol and data representation. However, this system requires that the datasets be converted into a specific internal representation. Moreover, integrating a new data source or client requires significant programming effort.

At Ohio State, we have proposed the notion of *automatic data virtualization* [27,36] and *automatic wrapper generation* [39], and have been developing grid middleware to support these. In both these approaches, a declarative description of a data format (a *layout descriptor*) is stored as part of the metadata associated with the data. In the data virtualization approach, an application is developed assuming a high-level or virtual view of the dataset, such as a relational table or an XML dataset. Using the layout descriptor, a middleware layer executes the application on the low-level dataset. In the automatic wrapper generation approach, a wrapper program is generated automatically using layout descriptors corresponding to both the source and the target formats.

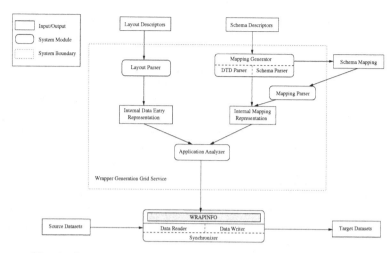

Fig. 2. Overview of the Automatic Wrapper Generation System

These approaches offer many advantages over the existing systems. New data sources and changes to existing data sources can be accommodated by just writing or modifying the layout descriptor associated with that source. Similarly, new analysis programs or updates to these programs can be easily incorporated as part of an application. Furthermore, both data sources and analysis tools can be discovered *on-the-fly*, and integrated with other sources or tools. This approach is also very efficient, as unnecessary data transformations are avoided.

We now briefly describe our tool for automatic wrapper generation. We use a layout descriptor for describing the format for each resource. Our layout descriptor is similar in flavor to the Data Format Definition Language being developed by the DFDL Working Group in the Global Grid Forum[2]. This allows us to integrate our work with the Data Access and Integration Standards (DAIS) proposed by the global grid forum. Such descriptions provide sufficient information for the system to understand the layout of binary or character flat-files, without relying on any domain- or format-specific information. Based on such information, our system is able to discover and transform information from one low-level dataset to another low-level dataset. This approach can efficiently transform large volumes of data between a large number of sources, while avoiding unnecessary data transformations and tedious maintenance of manually written wrappers. For each resource, only one layout descriptor needs to be written for each of its input and output formats. Moreover, as new data sources or tools are published, or move to a new format, only their layout descriptors needs to be written or rewritten.

In order to generate a wrapper that is capable of transforming a dataset of a general format into another dataset of a general format, the system needs to have information about the physical data layouts. It also needs to understand the user's logical view of the data (i.e. the schema) so that it can draw the correspondence between the input and output datasets. We have designed a

---

[2] Please see http://forge.gridforum.org/projects/dfdl-wg

layout description language to achieve both of the above. The information about both the source and the target data layouts are represented using our layout description language. Tabular-structured input or output schemas can also be described using the same language, whereas semi-structured input or output schema are described using the XML DTD format. The *layout parser* parses the layout descriptors and generates internal data entry representations. The schemas are input into the *mapping generator*, which generates the mapping between the source and the target data schema. The inferred schema mapping is presented to the user in a flat file so that it can be verified or modified if needed.

The internal representation of data entries and the mapping completely defines a wrapping task and the functionality of a wrapper can be inferred from them. This inference can be carried out by either the wrapper generation system, or the wrapper itself. For a better overall system performance, we need to reduce the computations performed by the wrapper, and also allow it to execute independent of the wrapper generation system. Therefore, a wrapper generation system module, *Application Analyzer*, performs all the analysis and summarizes important application-specific information for the wrapper in a data structure, which we refer to as the WRAPINFO data structure.

The wrappers work independently from the wrapper generation system. Our wrappers comprise three modules, the DataReader, the DataWriter and the Synchronizer, each of which is independent of the specific transformation task that needs to be carried out. The information specific to a wrapping task is already captured in the WRAPINFO data structure. Using this data structure as the input, these three modules can carry out a transformation task. The *DataReader* and the *DataWriter*, as their names suggest, are responsible for parsing the input dataset and writing to the output files, respectively. The *Synchronizer* serves as a coordinator between these two modules, as it forwards the values constructed by the DataReader to the DataWriter, and manages the input dataset buffer.

Our design is very well suited for generating wrappers to carry out transformation tasks in a grid environment. Wrapper generation can be easily implemented as a grid service. As shown in the preliminary experimental evaluation of this system (see [39]), for large datasets, the wrapper generation time is a very small fraction of the actual wrapper execution time. The wrapper generator only requires the layout descriptors as input. In comparison, a wrapper needs to be executed at a location where the data movement costs for the input and output datasets are minimized. At the same time, the transformation time can be high, and the wrapper needs to be executed efficiently. By designing the wrapper with application independent modules and representing the WRAPINFO data structure in a machine independent XML file, we make it simpler for the wrappers to be ported for efficient execution on a variety of platforms.

2.3 FREERIDE-G: Middleware for Scalable Processing of Remote Datasets

Many challenging applications, including those in the environmental area, involve analysis of data from distributed datasets. This often leads to the need for *remote processing* of a dataset. Even if all data is available at a single repository, it is not possible to perform all analysis at the site hosting such a shared repository. Networking and storage limitations make it impossible to down-load all data at a single site before processing. Thus, an application that processes data from a remote repository needs to be broken into several stages, including a data retrieval task at the data repository, a data movement task, and a data processing task at a computing site.

An important challenge in this area, which we believe has received only a limited attention, is that careful coordination of storage, computing, and networking resources is required for efficiently analyzing datasets stored in remote repositories. Because of the volume of data that is involved and the amount of processing, it is desirable that both the data repository and computing site may be clusters. This can further complicate the development of such data processing applications.

We have been developing a middleware, FREERIDE-G (FRamework for Rapid Implementation of Datamining Engines in Grid), which supports a high-level interface for developing data mining and scientific data processing applications that involve data stored in remote repositories. Particularly, we have the following two goals behind designing the FREERIDE-G middleware:

Support High-End Processing: Parallel configurations, including clusters, are being used to support large scale data repositories. Many data mining applications involve very large datasets. At the same time, data mining tasks are often compute-intensive, and parallel computing can be effectively used to speed them up [38]. Thus, an important goal of the FREERIDE-G system is to enable efficient processing of large scale data mining computations. It supports use of parallel configurations for both hosting the data and processing it.

Ease Use of Parallel and Distributed Configurations: Developing parallel data mining applications can be a challenging task. In a grid environment, resources may be discovered dynamically, which means that a parallel application should be able to execute on a variety of parallel systems. Thus, one of the goals of the FREERIDE-G system is to support execution on distributed memory and shared memory systems, as well as on cluster of SMPs, starting from a common high-level interface. Another major difficulty in developing applications that involve remote data is appropriate staging of remote data, and possibly caching when feasible and appropriate. FREERIDE-G is designed to make data movement and caching transparent to application developers.

Prior Work: FREERIDE Middleware. Our proposed work on FREERIDE-G is based on an earlier system, FREERIDE (FRamework for Rapid Implementation of Datamining Engines) [20,21]. This system was motivated by the difficulties in implementing and performance tuning parallel versions of data mining algorithms. FREERIDE is based upon the observation that parallel versions of several well-known data mining techniques share a relatively similar structure. We have carefully studied parallel versions of apriori association mining [1], Bayesian network for classification [3], k-means clustering [19], k-nearest neighbor classifier [17], artificial neural networks [17], and decision tree classifiers [29]. In each of these methods, parallelization can be done by dividing the data instances (or records or transactions) among the nodes. The computation on each node involves reading the data instances in an arbitrary order, processing each data instance, and performing a *local reduction*. The reduction involves only commutative and associative operations, which means the result is independent of the order in which the data instances are processed. After the local reduction on each node, a *global reduction* is performed. FREERIDE exploits this commonality to support a high-level interface for developing parallel implementations. The target environment are clusters of SMPs, which have emerged as a cost-effective, flexible, and popular parallel processing configuration. Clusters of SMP workstations, where each workstation has an attached disk farm, offer both distributed memory or shared-nothing parallelism (across nodes of the cluster) and shared-memory parallelism (within a node).

FREERIDE has been successfully used for developing parallel versions of a number of common mining algorithms, including apriori and FP-tree based association mining, k-means and EM clustering, decision tree construction, and nearest neighbor search [14,20,21,22,23]. More recently, in collaboration with Raghu Machiraju's group at The Ohio State University, we have used FREERIDE for two scientific data analysis application, vortex analysis [16] and molecular defect detection [15,33]. Our experimental studies with both data mining and scientific data analysis applications have shown that FREERIDE allows very high parallel efficiency, allows scaling to disk-resident datasets, and simplifies parallel implementation tasks.

FREERIDE-G. FREERIDE-G system will be directly built on the FREERIDE middleware described in the previous section. The critical new functionality we have added in the middleware is as the ability to support mining and processing of data from remote repositories.

Figure 3 shows the three major components: (the *data server*, the *resource selection framework*, and the *compute node client*). The data server runs on every on-line data repository node in order to automate data delivery to the end-users

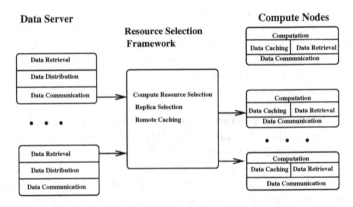

Fig. 3. FREERIDE-G System Architecture

processing node(s). The resource selection framework has the following goals: 1) Finding computing resources, 2) Choosing replica, and 3) Finding non-local caching resources. A compute server runs on every end-user processing node in order to receive the data from the storage repository and perform application specific analysis of it. In our ongoing work, we are integrating the resource selection framework of FREERIDE-G with WS-Resource Framework.

3 Data Analysis Services

In this section, we describe the data analysis services we will be developing on top of the middleware components described in the previous section. Our goal will be to develop these independent of the specific datasets on which they may be applied, or the resources on which they may be executed. Thus, these services will be a part of the cyberinfrastructure for environmental applications.

In anticipation of the test cases to be deployed here, the services will fall into three categories: Multi- Model/Data Integration Services, Querying Services, and Mining Services. In surveying the various applications these services address, it is noticed that there are two extremes of accessibility and usability possible with these services: 1) Applications that involve quick, near real-time forecasting/nowcasting problems such as coastal hazards and environmental disruptions; and 2) Applications to slower time base problems such as coastline and global climate change. The former relies heavily on the simultaneous integration of three dimensional physics-based models and one and two dimensional data sets in streaming conditions to achieve timely accurate predictions. The latter relies most heavily on the fusion of one, two, three, and four dimensional datasets, mostly from sensors to determine highly precise changes, such as those in geometric coastal aspects such as shoreline position and erosion, or environmental quality trends required for global climate change assessment.

3.1 Multi-model Multi-sensor Data Integration Service

Three-dimensional data mapping from multi-sensor images, which are taken at different times and altitudes and with different imaging principles (frame or pushbroom), is a challenging task. Based on novel techniques developed under Li's group at The Ohio State University, we will be developing a grid service for three-dimensional multi-sensor data integration. This service will be built on top of our middleware for wrapper generation, and will be widely applicable to a number of environmental applications.

This service will include three sub-service modules: Universal spatial data transfer and conversion service, Multi-model integration service, and Multi-sensor 3-D mapping service.

Universal Spatial Data Transfer and Conversion Service. Multidimensional data used in the environmental research is heterogeneous and there are different data types, data formats, reference systems, spatial resolutions, time scales, and accuracies associated with the data. For example, for coastal informatics, vertical datums of most water-side data are based on tidal datums such as Mean Sea Level (MSL), while those of land-side data are based on Geoid models such as the North American Vertical Datum 1988 (NAVD88). Conversions between various vertical datums are needed for integrating data on both water side and land side [7].

Universal spatial data transfer and conversion (USDTC) service will be based on the Wrapper Middleware described in the previous section. This will be a two-step service, involving data format conversion and reference system conversion. By using our wrapper generation service, user needs to input the description of the data format and URL of the data to be converted. We will integrate our layout description language with GML, which is an XML encoding for the transport and storage of geographic information, including both the geometry and properties of geographic features[3]. As needed, reference system conversion service will convert the spatial data into a predefined common georeference system. For example, for the spatial data in Lake Erie, the vertical datum will be NAVD88 and the horizontal datum will be the North American Datum (NAD83) with an UTM coordinate system in meters.

Multi-Model Integration Service. There exist several coastal modeling and observing systems specifically designed for different coastal areas. However, there is very little direct coupling between these in terms of sharing data and functionalities. There is a clear need for a sustained and Integrated Ocean Observing System (IOOS) that will make more effective use of existing resources, new knowledge, and advances in technology for ocean and coastal management. Multi-model integration service will be designed for integrating different coastal modeling and observing systems. The integration of different systems is realized by controlling the input/output formats of the systems through the universal spatial data transfer and conversion service. All communications between different modeling and observing systems and databases go through the wrapper middleware. For example, when Model 1 requests data from Database 1, it will submit a request to Database 1 along with the required data description. Once Database 1 receives the request, it will extract the requested data, conduct necessary data format conversion using wrapper middleware and reference system conversion. Finally, the wrapper middleware parses the converted GML to the data format requested by Model 1.

Multi-sensor 3-D Mapping Service. Stereographic mapping using aerial photographs is a mature technology for topographic mapping. 3-D mapping using high-resolution satellite (IKONOS and QuickBird) images has been researched in recent years and accuracies comparable to those from aerial images have been reported [24,31,32,37]. A systematic study of the 3-D accuracy improvement of IKONOS imagery with various adjustment models was conducted and a number of practical guidelines have been presented [18]. However, 3-D mapping from multi-sensor images, which are taken at different times and

---

[3] http://www.opengis.org

altitudes and with different imaging principles (frame or pushbroom), is an unsolved and challenging task.

We will develop a comprehensive geometric model that depicts the transformation between ground point and image point. Two types of rigorous sensor/-camera models, a frame camera model (for aerial photographs) and a generic linear scanning (pushbroom) sensor model (for satellite images), along with one replacement sensor model, namely a rational functional model (for satellite images and aerial photographs whose rigorous sensor model is not released, e.g., IKONOS images [25]), will be incorporated into the comprehensive model. With sufficient ground control points, the sensor model parameters are solved/refined in a unified least-squares adjustment. The parameters to be solved/refined include the exterior orientation parameters (position of the optical center and three rotation angles) of the frame camera model, polynomial coefficients of the pushbroom model that depict the change of exterior orientation parameters along scan lines, and additional parameters in image space to refine the rational functional model. Using the least-squares adjustment, this will provide the optimal orientation parameters of the multi-sensor images, and high accuracy 3-D positions of the tie points that link all the images together. In addition to this comprehensive model, a systematic study will be conducted to evaluate the best geometric configuration and to select the best combination of images based on the analysis of altitude, baseline, convergent angle, and error propagation.

Based on the adjusted sensor orientation parameters and image matching technology, a seamless digital terrain model (DTM) can be generated. An innovative image matching method, based on the vision invariance theory, a coarse-to-fine strategy, and epipolar geometry, will be developed for multi-sensor image matching. Available rough DTM can be used in image matching to limit the search range and thus to improve matching reliability. Subsequently, quantitative information about 3-D shoreline positions can be derived with a high level of accuracy using semi-automatic methods.

3.2 Querying Service

The second service we will provide will be for planning and processing queries across environmental data repositories. Overall, our goal will be to support efficient and unified access over heterogeneous, high-dimensional spatio-temporal coastal data. This will include several components: extracting suitable metadata for image datasets and managing them using existing metadata catalogs [6,34], query planning using such metadata, and query execution using our support for data virtualization.

The distributed coastal data repositories involve large amounts of spatio-temporal data in the form of sequences of images and dynamic streams of empirical measurements, generated by a diverse set of data sources. The spatio-temporal data is typically multi-dimensional, i.e., they have multiple attributes that may be queried and analyzed, and/or include high-dimensional semantic representations of aerial and satellite images, raster maps, and digital terrain models. To enable query planning, a metadata catalog summarizing the contents of each data source is required. While storing, discovering, and accessing metadata has been widely studied, choosing appropriate metadata for image databases is an open problem. The image sources are inherently large and each image will be represented by multi-dimensional metadata describing the low-level and high-level features of the image. Images from satellites, including SeaWiFS, MODIS, Landsat 7 ETM+, ASTER, Envisat, IKONOS, and QuickBird, are being processed at the OSU Mapping and GIS Lab directed by Professor Li, a co-PI for this project. Software developed at the Lab will be used, along with commercial software such as Erdas Imagine and PCI Geomatics, to perform comprehensive operations such as image enhancement, geo-registration, map projection, image classification, feature extraction, coastal object recognition, raster to vector conversion, image matching, DEM and orthophoto generation. Shorelines, wetland boundaries, sea grasses sediment, and land cover information are extracted automatically and/or semi-automatically and converted to metadata, e.g., high-dimensional feature vectors. To generate a metadata index, we have started exploring multi-stage VQ and split-VQ, which are well suited for limited memory applications by imposing certain structural constraints on the codebook. Our preliminary experiments on satellite images establish that both split-VQ and multi-stage VQ can be effectively employed to design negligibly small codebooks for a large-scale image database [35]. We will implement a two-level VQ, which we refer to as Split-MS VQ, to index the distributed image repositories on the Grid.

Query submission and composition will be handled by the central server, which also performs query planning and routing of sub-queries to the necessary database servers. An overall schema for the underlying distributed data sources will be defined in the querying service tool along with a query planner module that selects the necessary paths to execute a given query. For efficient access, the query planner will take into consideration characteristics of the participating nodes, connectivity between each node, and query characteristics, i.e., what data is being requested and where clients reside. Access frequencies of the data can be used to generate better query plans over heterogeneous nodes on the Grid. Given a potential query workload and varying cost of storage and retrieval, we plan to address the following questions: how should one create a metadata lookup table and index structure for the distributed spatio-temporal and image data, how much replication, if any, should one use, and what strategies should be used to replicate this data.

3.3   Mining Services for Heterogeneous and High-Dimensional Environmental Data

Environmental applications broadly involve two types of data analysis tasks. The first type involve one or more large databases, and scalable and *remote* analysis is desired. The second type involves real-time analysis of distributed streaming data. As part of our cyberinfrastructure component for environmental applications, we will develop a number of grid mining services, using novel algorithms being developed by Ferhatosmanoglu's group. The first type of analysis tasks will be implemented using the FREERIDE-G middleware described in the previous section, and the second type of analysis tasks will be implemented using GATES middleware. Both these systems have been developed by Agrawal's group, and were described in the previous section. All services will be coupled with the wrapper generation tool, to allow the algorithm implementations to be independent of the specific layout of a sensor data source or database.

Scalable and Remote Mining Tasks. The distributed coastal environmental data demonstrate a great example of database heterogeneity in both space and time. In addition to the image data, another typical data type is high-dimensional time series, including continuous logging of water-level and tide observations at various gauge stations and buoys at fine time intervals (equal or non equal), which have been collected for decades. Finding specific water-level and tidal data relevant to specific locations and associated image data within the large distributed coastal database poses a great challenge. The current time series data mining methods generally assume that the data are collected in equal length intervals, and that in comparing time series, the lengths are equal to each other. However, in environmental monitoring applications, the length of observations are neither fixed nor standardized, and the data sources are different (e.g., waves, currents, wind, altimetric readings, etc.). Approaches for mining time series data need to be revisited keeping the heterogeneity and wide range of requirements in mind. A straightforward data mining process would apply the algorithms globally over the whole data set. For example, to identify strongly related groups of factors affecting the environment, one can apply a standard clustering algorithm, where coastal attributes, observed by various data collection technologies, are clustered using a distance measure defined over the corresponding high-dimensional vectors. However, the result of such a process would have little or no meaning partially because of the obvious problems caused by high-dimensionality, heterogeneity and incompleteness of the data, and partially because of the difficulty of interpreting the output of such an analysis.

Since global mining of complex coastal databases is infeasible, we have been developing an *information-mining* approach consisting of a simple preprocessing, with minimal assumptions on the data, followed by two major steps [2]. In the first step, significant and homogeneous subsets of data (e.g., data generated by similar sources) are selected and analyzed using the mining algorithm of interest. In the second step, the information gathered in the first step is joined by identifying common (or distinct) patterns over the results of mining of the subsets. We plan to apply this framework to a variety of data mining techniques over heterogeneous and high-dimensional environmental data sequences. For example, for clustering, the first step corresponds to the clustering of sensors for each data collection campaign, and the second step corresponds to finding strongly related groups of sensors (or a set of representative sensors with minimal cross-correlations) by mining common patterns over the clusters generated in the first step. This algorithm will have three distinct results: a) groups of sensors, data sources, or data types that detect relevant coastal environmental changes, b) a global panel of sensors or data attributes, with corresponding weights, that can model the coastal impacts, and c) the distinct patterns of outliers and adverse events which provide valuable information for the abnormal status of the coastal environment and the ecosystem.

It is clear that any single clustering method or any single distance metric would not be enough to capture all types of relationships, especially on such a non-standard and high-dimensional set of data. The proposed two-step process can be used to pool the information from different metrics, different clustering algorithms, and from different sources of environmental data, resulting in more robust outcomes. It minimizes the differences among time series caused by source variation, hence, local groups of time series become equal length and equal interval, which makes many powerful distance metrics applicable. The framework can also be employed in collaborative studies, where the information rather than the data across multiple sources is pooled and analyzed to extract knowledge that may not be derivable using a single source. Sequence data mining and clustering have been extensively studied from the perspective of efficiently searching or extracting rules from them. A similar approach can be developed for the proposed system to see if any of the factors has an obvious effect on the model. However, the current techniques will not capture multi-variate relationships between involved attributes and factors analyzed. We plan to develop similarity measures and models even when the number of observations per each attribute is relatively short, which is the case for many similar information systems applications.

FREERIDE-G middleware, which has already been used for creating grid-based and scalable implementations of clustering algorithms like k-means and EM, will be used for implementing these algorithms. This will create services which can flexibly use parallel environments for scalable execution, and will perform data retrieval and data transfers transparently in a grid environment.

Managing and Mining Real-time Environmental Data Streams. Real-time environmental monitoring requires continuous querying and analysis over distributed data streams collected by dynamic data acquisition technologies, such as land- and water-based in situ sensors. We are particularly interested in enabling ways to utilize the data repository to monitor and characterize chemical, hydrological, thermal, and seismic changes over both space and time. The concept of data streams has recently attracted considerable attention in database and networking communities. A fundamental challenge in distributed data stream management is to develop an online and distributed compression algorithm to minimize the cost of communication among distributed nodes of the Grid. The resulting compressed data can also be utilized, centrally or in a distributed fashion, by many data mining tasks such as clustering, classification, change detection, statistical monitoring, selectivity estimation, query optimization, and efficient query processing over multiple streams. For example, continuous similarity queries can be executed to detect weather patterns similar to a previously known pattern. Similarity joins over streaming and archival data can be used in modeling and classifying a new object as a certain terrain property, by comparing its attributes to a set of previously classified objects in the archival database. Correlation-based joins can be implemented over the summaries to identify pairs of similar streams such as water-level and quality observations.

Following the success of quantization-based approaches for querying large-scale and high-dimensional data repositories, we have proposed a scalar quantizer, i.e., independent quantization, to compress multiple data streams [28]. Each newly arrived set of data elements is quantized without accessing previously summarized data, which makes independent quantization particularly appropriate for dynamic and distributed data streams. Independent quantization suffers from the fact that correlation between data elements is partially ignored in return for efficiency. Removal of redundancy via signal transforms followed by quantization is known to be more effective than purely quantization-based or purely transform-based strategies [13]. We successfully applied this two step process to static high-dimensional databases [8], where preprocessing of data is amortized with gains in query processing. However, it is not of immediate use as an online technique, because such preprocessing is infeasible on real-time streaming data. A technique is needed that would achieve the best of both worlds, i.e., that would be as effective as transform domain processing in removing redundancy and as fast as independent quantization for online computation of the summary. Investigation of prediction-based methods will be a valuable step since the coastal data of practical interest has a time-varying nature and possess a high amount of correlation between nearby sample values. As a first step, we are developing an online predictive quantizer where the correlation between the data elements is exploited through the prediction of the set of incoming elements of a correlated group of streams in terms of th latest few elements. In this prediction, the error is efficiently quantized at each instant independently by utilizing a different setting for that instant. In traditional applications, a prediction coefficient is used over a single signal, which is assumed to be stationary. The overhead of the coefficient is amortized by using the same coefficient over a long period of time. For multiple streams in environmental monitoring applications, the same coefficient can be utilized for a group of correlated data sources, such as spatially close sensors. The coefficient is amortized over multiple streams, and can be updated for each time instant for non-stationary signals. We have tested a preliminary version of this approach over weather data streams obtained from the National Climatic Data Center. The predictive quantization based approach significantly outperforms current transform-based techniques [30] both in data representation quality and in precision and recall for various types of queries. Even when the current approaches use extra preprocessing over the data, this approach achieves better results with its online algorithm. An unoptimized implementation was capable of handling more than 600,000 streams per second on a modest PC.

The above techniques will be implemented using GATES middleware, which will result in widely deployable grid-services that can be executed on distributed streaming sources.

4 Applications

The two end applications where the cyberinfrastructure component will be deployed are: real-time coastal forecasting and nowcasting, and longer-term coastal erosion analysis and prediction. As part of this project, we are proposing to develop and test end-to-end systems for these two applications, using real data from a number of sources.

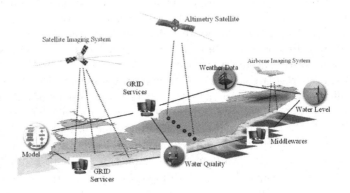

Fig. 4. Data Collection in Lake Erie

The coast is an area of intensive interaction between land, sea, and air. More than 80 percent of the U.S. population lives within 50 miles of the coast [26]. This coastal zone is exploited by humans for food, recreation, transport, waste disposal, and other needs. However, excessive discharge and uncontrolled human activities, in addition to natural processes, have created environmental problems in the coastal zone such as habitat modification, habitat destruction, and ocean pollution.

Our deployment and evaluation will be carried out in context of the GLOS, the Great Lakes Observation System[4]. GLOS already has a full three dimensional nowcast and forecast system, which include GLFS: Great Lakes Forecasting System operated at The Ohio State University, and GLCFS: Great Lakes Coastal Forecasting System operated by the NOAA Great Lakes Environmental Research Laboratory, NOAA National Ocean Service, and NOAA National Weather Service. This system makes forecasts for all five Lakes, including real time predictions of waves, water levels, and 3-D currents, and temperature and salinity fields. Forecasts are made twice daily for a 36 hour range and nowcasts are made every hour. This system has been operational since 1990. Therefore, the GLOS now consists of permanent in situ collection data, remotely acquired data of all types and wavelengths, and a fully functioning nowcast/forecast system.

---

[4] http://glos.us

4.1   Real-Time Coastal Nowcast and Forecasts

This application study will focus on creating a distributed, real-time cyberinfrastructure application, using the GATES middleware and associated services for mining time-series data.

Background: Nowcast/Forecast (N/F) systems contain coupled subsystems of two and three dimensional models that are continuously exchanging data, sensor data input consisting of continuous, streaming inputs of time series of data at fixed points, and periodic 2 dimensional data sets from remote satellites. The existing nowcast system for Lake Erie (Michigan) consists of inputs of real time series of weather data collected at fixed points around the Lake by two different agencies (Canadian and NWS), fixed point weather time series' data collected by the Coast Guard, weather data collected by NOAA Ships Observations Program, which collects data from ships underway across the particular Lake, in situ weather and temperature buoy data collected jointly by NOAA and Canada at four fixed sites in Lake Erie, and water-level time series' collected at 13 fixed points around the Lake. Finally, real-time tributary input flow data are collected for the four most important tributaries in the Lake from USGS. We also use an adhoc basis two channels of AVHRR infrared and visible band satellite data to get temperature fields at 1 km^2 resolution and the same two bands of GOES-8 data to help with estimation of cloud cover for the heat flux calculations.

To assess the nowcasts, comparisons are made between the predictions made at a certain hour to data collected for that particular hour as measured by the same suite of data summarized above. That is, if we make a nowcast estimate valid at 1:00PM (typically at 1:40 by the time all components are completed) we will make an assessment of its accuracy by comparing the estimate to the actual data measured precisely at 1:00 PM. To perform forecasts, we need to take short range weather forecast data predicted by NWS, extract the gridded fields that are available from the models at a height above the water surface, regrid them to match the grid on the Lake N/F system, adjust the fields so they are valid at the Lake surface, and the use them as input to the N/F models. This process takes a bit longer than the process required for just the nowcasts.

The data collection environment in Lake Erie, for both real-time forecasts/nowcasts, and longer term coastal analysis is shown in Figure 4.

Problems and Approach: There are three areas of concern in the present GLFS and GLCFS that we believe our cyberinfrastructure components can improve. First, the systems as they stand now are completely *hardwired*. For the most part, they closed to external coupling or cooperation. This was initially due to the need to get to operational status and the lack of sound networking tools back in the early to mid 1990s', when the first system was assembled. Therefore if a researcher wants to attach and test their own new forecast algorithm that extends the system, say for example, to beach closings and coliform forecasts, they would have to stop the system, and completely rebuild the system with the new data feeds and code in it. This is clearly not desirable from the view-point of advancing the system and involving the research community.

Second, the hardwired nature of the system also prevents new data, which may be collected either on an ad-hoc research basis or incrementally, to be ingested into the system. Such might be the case when a broad field program provides a suite of very rich data, say, for a period of two years. Finally, the hardwired nature of the present system prevents relative ease of extracting and interpreting (querying, mining) data that are being predicted while the N/F are being made.

This application study using cyberinfrastructure components will focus on improvements in each of these three areas. We will construct incrementally more robust test cases over the period of the project.

4.2 Coastal Erosion Prediction and Analysis

The second application will focus on coastal change monitoring and modeling. This important and challenging application will serve as a demonstration of the proposed data integration services, query services, and data mining services. Ohio shore along Lake Erie will be use as a test site.

Background: Erosion along the Ohio shore of Lake Erie is a serious problem. Many beaches along Ohio's lakeshore were eroded due to the record-high lake levels during 1970s - 1990s. The leftover unprotected bluffs, especially in the Lake County east of Cleveland, OH, are even more vulnerable to wave erosion. Each year, nearly 1.6 million tons of material is eroded, threatening public safety, health, and welfare and causing extensive damage to residential, commercial, industrial, and agricultural property. Economic losses caused by coastal erosion in the Great Lakes region were estimated at 290 million in 1985 and 1986 and at 9 million in 1985 in Lake County, Ohio (NOAA and ODNR, 1999).

Increased attention and effective action has been taken by the federal and state agencies. The Ohio Department of Natural Resources (ODNR) identifies Coastal Erosion as one of several priority coastal management issues (NOAA and ODNR, 1999). To minimize coastal erosion damages, ODNR was directed to identify coastal erosion areas along Lake Erie shore. Accurate coastal change information is crucial to identification of erosion areas as well as to many other coastal applications, including coastal development, coastal environmental protection, and coastal resource management, and decision making.

Advances in technologies, including space- and airborne imaging and land- and water-based in situ sensors have been extensively utilized in coastal research and applications. Large quantities of the coastal variation information are being used to characterize the coastal physical environment with macro and micro changes in both space and time. It is essential to integrate multiple sensor data in order to best achieve the coastal change monitoring results that may not be achievable using any single sensor.

Remote sensing technologies have been widely used as a major data source for coastal monitoring, planning, conservation, evaluation, and management. Coastal information such as coastal land use/land coverage, water quality, habitat distribution, 3-D shoreline, coastal digital terrain model, etc., can be extracted from remote sensing data by applying certain image classification and feature extraction algorithms. Coastal change information can also be detected by comparing the above extracted information to the corresponding historical one. Sometimes, historical imagery also needs to be processed to extract coastal information which is not available in coastal data repositories.

Problems and Approach: In general, historical coastal data are located at different federal, state, and local agencies, or research institutes. In order to perform coastal change analysis, traditionally we have to request and collect those data from different places and process them in a local computer. Due to the bandwidth and storage limits, only those data in a small area can be obtained and processed. In case we need to extend the study area, all the steps of data collection and processing have to be repeated. Such kind of time-consuming and data-duplication situations arise very often in current coastal research activities.

Another drawback of current approaches is that this type of operation often causes delay between the time when data are available from the data provider, and the time when user receives the updated news about the data availability. Such delay can not ensure data synchronization in coastal change analysis. Thus, inaccurate and untimely results have become inevitable.

In our cyberinfrastructure design, and through the use of the middleware systems developed here, multiple-sensor integration will be realized in a highly accurate, seamless, and systematic way. This will ensure high quality of the integrated data for coastal change monitoring and modeling. Based on the previous and current research results of Ron Li, a co-PI of this proposal, on coastal change analysis, data fusion, and shoreline change modeling, we will be deploying the cyberinfrastructure component for coastal change monitoring, analysis, and modeling.

This system will include the following two components: 1) Coastal information extraction and change detection, and 2) Coastal change and erosion pattern analysis for delineation of coastal erosion areas. The proposed multi-model multi-sensor integration service will serve as the backbone of the system. It will support communication between distributed coastal data and functions as well as necessary conversions of data formats and reference systems. The multi-sensor 3-D mapping service will be used to process IKONOS and QuickBird imagery for extracting 3-D coastal features such as 3-D shoreline and DEM. The query services will be used to retrieve historical coastal information for detection change information.

We will use the query service and data mining services to perform coastal change and erosion pattern analysis. Statistical analyses of shoreline changes will be conducted to find out the erosion causes. The analyses include: 1) Correlation analysis between the coastal shoreline position change, land use, and water levels, 2) Correlation analysis between the coastal change and human activities such as coastal protection, and 3) Severe erosion pattern analysis caused by consistent natural processes, human activities, or both. Our proposed mining services will be used for each of these.

5   Conclusions

This paper has given an overview of a recently initiated cyberinfrastructure project at The Ohio State University. This project proposes to develop and evaluate a cyberinfrastructure component for environmental applications. This will include developments in middleware, model integration, analysis, and mining techniques, and the use of a service model for supporting two closely related applications. These applications will be real-time coastal nowcasting and forecasting, and long-term coastal erosion analysis and prediction.

References

1. Agrawal, R., Shafer, J.: Parallel mining of association rules. IEEE Transactions on Knowledge and Data Engineering 8(6), 962–969 (1996)
2. Altiparmak, F., Ferhatosmanoglu, H., Erdal, S., Trost, C.: Information mining over heterogeneous and high dimensional time series data in clinical trials databases. In: IEEE Transactions on Information Technology in Biomedicine
3. Cheeseman, P., Stutz, J.: Bayesian classification (autoclass): Theory and practice. In: Advanced in Knowledge Discovery and Data Mining, pp. 61–83. AAAI Press / MIT Press (1996)
4. Chen, L., Reddy, K., Agrawal, G.: GATES: A Grid-Based Middleware for Distributed Processing of Data Streams. In: Proceedings of IEEE Conference on High Performance Distributed Computing (HPDC). IEEE Computer Society Press, Los Alamitos (2004)
5. Cornillon, P., Callagher, J., Sgouros, T.: OpeNDAP: Accessing data in a distributed, heterogeneous environment. Data Science Journal 2, 164–174 (2003)
6. Deelman, E., Singh, G., Atkinson, M.P., Chervenak, A., Hong, N.P.C., Kesselman, C., Patil, S., Pearlman, L., Su, M.: Grid-Based Metadata Services. In: Proceedings of the 16th International Conference on Scientific and Statistical Database Management (SSDBM 2004) (2004)
7. Zhou, F., Niu, X., Li, R.: Vertical datum conversion in lake erie: A technical paper. Voice of the Pacific, PACON Newsletter (2003)
8. Ferhatosmanoglu, H., Tuncel, E., Agrawal, D., El Abbadi, A.: Vector approximation based indexing for non-uniform high dimensional data sets. In: Proceedings of the 9th ACM Int. Conf. on Information and Knowledge Management, McLean, Virginia, pp. 202–209 (November 2000)
9. Ferris, C., Farrell, J.: What are Web Services. In: Communications of the ACM (CACM), pp. 31–35 (June 2003)
10. Foster, I., Kesselman, C., Nick, J., Tuecke, S.: Grid Services for Distributed Systems Integration. IEEE Computer (2002)
11. Foster, I., Kesselman, C., Nick, J.M., Tuecke, S.: The Physiology of the Grid: An Open Grid Services Architecture for Distributed Systems Integration. In: Open Grid Service Infrastructure Working Group, Global Grid Forum (June 2002)
12. Foster, I., Kesselman, C., Tuecke, S.: The Anatomy of Grid: Enabling Scalable Virtual Organizations. International Journal of Supercomputing Applications (2001)
13. Gersho, A.: Vector Quantization and Signal Compression. Kluwer Academic Publishers, Dordrecht (1992)
14. Glimcher, L., Agrawal, G.: Parallelizing EM Clustering Algorithm on a Cluster of SMPs. In: Proceedings of Europar (2004)

15. Glimcher, L., Agrawal, G., Mehta, S., Jin, R., Machiraju, R.: Parallelizing a Defect Detection and Categorization Application. In: Proceedings of the International Parallel and Distributed Processing Symposium (IPDPS) (2005)

16. Glimcher, L., Zhang, X., Agrawal, G.: Scaling and Parallelizing a Scientific Feature Mining Application Using a Cluster Middleware. In: Proceedings of the International Parallel and Distributed Processing Symposium (IPDPS) (2004)

17. Han, J., Kamber, M.: Data Mining: Concepts and Techniques. Morgan Kaufmann Publishers, San Francisco (2000)

18. Wang, J., Di, K., LI, R.: Evaluation and improvement of geopositioning accuracy of ikonos stereo imagery. Journal of Surveying Engineering (2005)

19. Jain, A.K., Dubes, R.C.: Algorithms for Clustering Data. Prentice-Hall, Englewood Cliffs (1988)

20. Jin, R., Agrawal, G.: A middleware for developing parallel data mining implementations. In: Proceedings of the first SIAM conference on Data Mining (April 2001)

21. Jin, R., Agrawal, G.: Shared Memory Parallelization of Data Mining Algorithms: Techniques, Programming Interface, and Performance. In: Proceedings of the second SIAM conference on Data Mining (April 2002)

22. Jin, R., Agrawal, G.: Communication and Memory Efficient Parallel Decision Tree Construction. In: Proceedings of Third SIAM Conference on Data Mining (May 2003)

23. Jin, R., Agrawal, G.: Shared Memory Parallelization of Data Mining Algorithms: Techniques, Programming Interface, and Performance. In: IEEE Transactions on Knowledge and Data Engineering (TKDE) (2005)

24. Di, K., Ma, R., Li, R.: Geometric processing of ikonos stereo imagery for coastal mapping applications. In: Photogrammetric Engineering and Remote Sensing (2003)

25. Di, K., Ma, R., Li, R.: Rational functions and potential for rigorous sensor model recovery. In: Photogrammetric Engineering and Remote Sensing (2003)

26. Mayer, L.A., Barbor, K.E., Boudreau, P.R., Chance, T.S., Fletcher, C.H., Greening, H., Li, R., Mason, C., Snow-Cotter, S., Wright, D.J., Lewis, R.S., Feary, D.A., Schaefer, T., Forsbergh, Y., Mason, B., Schrum, A.: A geospatial framework for the coastal zone. In: National Needs for Coastal Mapping and Charting (edited book) (2004)

27. Li, X., Agrawal, G.: Supporting XML-Based High-level Interfaces Through Compiler Technology. In: Rauchwerger, L. (ed.) LCPC 2003. LNCS, vol. 2958. Springer, Heidelberg (2004)

28. Liu, X., Ferhatosmanoglu, H.: Efficient k-nn search on streaming data series. In: International Symposium on Spatial and Temporal Databases, Santorini, Greece, pp. 83–101 (July 2003)

29. Murthy, S.K.: Automatic construction of decision trees from data: A multidisciplinary survey. Data Mining and Knowledge Discovery 2(4), 345–389 (1998)

30. Ogras, U., Ferhatosmanoglu, H.: Online summarization of dynamic time series data. VLDB Journal

31. Li, R., Di, K., Ma, R.: 3-d shoreline extraction from ikonos satellite imagery. Journal of Marine Geodesy (2002)

32. Li, R., Zhou, G., Schmidt, N.J., Fowler, C., Tuell, G.: Photogrammetric processing of high-resolution airborne and satellite linear array stereo images for mapping applications. International Journal of Remote Sensing (2002)

33. Mehta, S., Hazzard, K., Machiraju, R., Parthasarathy, S., Willkins, J.: Detection and Visualization of Anomalous Structures in Molecular Dynamics Simulation Data. In: IEEE Conference on Visualization (2004)
34. Singh, G., Bharathi, S., Chervenak, A., Deelman, E., Kesselman, C., Mahohar, M., Pail, S., Pearlman, L.: A Metadata Catalog Service for Data Intensive Applications. In: Proceedings of Supercomputing 2003 (SC 2003) (November 2003)
35. Tuncel, E., Ferhatosmanoglu, H., Rose, K.: VQ-index: An index structure for similarity searching in multimedia databases. In: ACM Multimedia, Juan Les Pins, France, pp. 543–552 (December 2002)
36. Weng, L., Agrawal, G., Catalyurek, U., Kurc, T., Narayanan, S., Saltz, J.: An Approach for Automatic Data Virtualization. In: Proceedings of the Conference on High Performance Distributed Computing (HPDC) (2004)
37. Niu, X., Di, K., Wang, J., Lee, J., Li, R.: Geometric modelling and photogrammetric processing of high-resolution satellite imagery. In: Proceedings of the XXth Congress of the International Society for Photogrammetry and Remote Sensing (ISPRS 2004) (2004)
38. Zaki, M.J.: Parallel and distributed association mining: A survey. IEEE Concurrency 7(4), 14–25 (1999)
39. Zhang, X., Agrawal, G.: Enabling information integration and workflows in a grid environment with automatic wrapper generation. In: SC 2005. ACM Press, New York (2005)

OGC® Sensor Web Enablement: Overview and High Level Architecture

Mike Botts[1], George Percivall[2], Carl Reed[3], and John Davidson[4]

[1] University of Alabama in Huntsville
mike.botts@uah.edu
[2] Executive Director, Interoperability Architecture
Open Geospatial Consortium (OGC), Inc.
gpercivall@opengeospatial.org
[3] CTO, Open Geospatial Consortium
creed@opengeospatial.org
[4] Image Matters, LLC
johnd@imagemattersllc.com

Abstract. The Open Geospatial Consortium (OGC) standards activities that focus on sensors and sensor networks comprise an OGC focus area known as Sensor Web Enablement (SWE). Readers interested in greater technical and architecture details can download the OGC SWE Architecture Discussion Paper titled "The OGC Sensor Web Enablement Architecture" (OGC document 06-021r1).

Keywords: Open Geospatial Consortium, Inc., OGC, sensors, sensor webs, standards, Sensor Web Enablement (SWE), Observations & Measurements Schema (O&M), Sensor Model Language (SensorML, Transducer Model Language (TransducerML or TML), Sensor Observations Service (SOS), Sensor Planning Service (SPS), Sensor Alert Service (SAS), Web Notification Services (WNS).

1 Introduction

Sensor Web Enablement (SWE) in the Open Geospatial Consortium, Inc. (OGC)[1] context refers to web accessible sensor networks and archived sensor data that can be discovered, accessed and, where applicable, controlled using open standard protocols and interfaces (APIs). Sensor location is usually a critical parameter for sensors on the Web, and the OGC sets geospatial industry standards, so SWE standards are being harmonized with other OGC standards for geospatial processing.

Members of the OGC are building a framework of open standards for exploiting Web-connected sensors and sensor systems of all types: flood gauges, air pollution monitors, stress gauges on bridges, mobile heart monitors, Webcams, satellite-borne earth imaging devices and countless other sensors and sensor systems.

---

[1] The OGC is an international consortium of industry, academic and government organizations who collaboratively develop open standards for geospatial and location services. (See http://www.opengeospatial.org.)

S. Nittel, A. Labrinidis, and A. Stefanidis (Eds.): GSN 2006, LNCS 4540, pp. 175–190, 2008.
© Springer-Verlag Berlin Heidelberg 2008

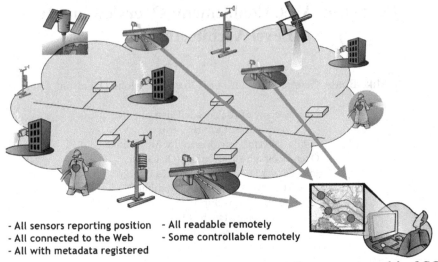

- All sensors reporting position   - All readable remotely
- All connected to the Web   - Some controllable remotely
- All with metadata registered

(Image courtesy of the OGC)

Fig. 1. Sensor Web Concept

SWE has extraordinary potential significance in many domains of activity, as the costs of sensor and network components fall, as their combined use spreads, and as the underlying Web services infrastructure becomes increasingly capable. The OGC consensus standards process coupled with strong international industry and government support in domains that depend on sensors has resulted in SWE standards that are quickly becoming established in all application areas where such standards are of use.

2 Overview

In much the same way that Hyper Text Markup Language (HTML) and Hypertext Transfer Protocol (HTTP) standards enable the exchange of almost any type of information on the Web, the OGC's SWE standards enable the Web-based discovery, exchange, and processing of sensor observations, as well as the tasking of sensor systems. The functionality includes:

- Discovery of sensor systems, observations, and observation processes that meet an application or users immediate needs;
- Determination of a sensor's capabilities and quality of measurements;
- Access to sensor parameters that automatically allow software to process and geo-locate observations;
- Retrieval of real-time or time-series observations and coverages in standard encodings

- Tasking of sensors to acquire observations of interest;
- Subscription to and publishing of alerts to be issued by sensors or sensor services based upon certain criteria.

Below is a list of the OpenGIS® Standards that make up the SWE suite of standards. Each specifies encodings for describing sensors and sensor observations and/or interface definitions for web services:

1. **Observations & Measurements Schema (O&M)** – (OGC Adopted Standard) Standard models and XML Schema for encoding observations and measurements from a sensor, both archived and real-time.
2. **Sensor Model Language (SensorML)** – (OGC Adopted Standard) Standard models and XML Schema for describing sensors systems and processes; provides information needed for discovery of sensors, location of sensor observations, processing of low-level sensor observations, and listing of taskable properties.
3. **Transducer Markup Language (TransducerML or TML)** – (OGC Adopted Standard) The conceptual model and XML Schema for describing transducers and supporting real-time streaming of data to and from sensor systems.
4. **Sensor Observations Service (SOS)** – (OGC Adopted Standard) Standard web service interface for requesting, filtering, and retrieving observations and sensor system information. This is the intermediary between a client and an observation repository or near real-time sensor channel.
5. **Sensor Planning Service (SPS)** – (OGC Adopted Standard) Standard web service interface for requesting user-driven acquisitions and observations. This is the intermediary between a client and a sensor collection management environment.
6. **Sensor Alert Service (SAS)** – (OGC Best Practices document) Standard web service interface for publishing and subscribing to alerts from sensors.
7. **Web Notification Services (WNS)** – (OGC Best Practices document) Standard web service interface for asynchronous delivery of messages or alerts from SAS and SPS web services and other elements of service workflows.

XML is a key part of the infrastructure that supports SWE. When the network connection for a sensor or system is layered with Internet and Web protocols, eXtensible Markup Language (XML) schemas defined in SWE standards can be used to publish formal descriptions of the sensor's capabilities, location, and interfaces. Then Web brokers, clients and servers can parse and interpret the XML data, enabling automated Web-based discovery of the existence of sensors and evaluation of their characteristics based on their published descriptions. The information provided also enables applications to geolocate and process sensor data without requiring *a priori* knowledge of the sensor system.

Information in the XML schema about a sensor's control interface enables automated communication with the sensor system for various purposes: to determine, for example, its state and location; to issue commands to the sensor or its platform; and, to access its stored or real-time data. This approach to sensor and data description also

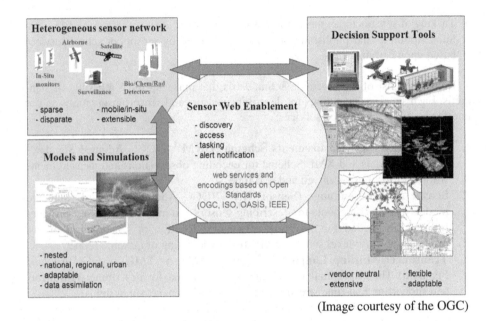

(Image courtesy of the OGC)

Fig. 2. The role of the Sensor Web Enablement framework

provides an efficient way to generate comprehensive standard-schema metadata for data produced by sensors, facilitating the discovery and interpretation of data in distributed archives.

3   The SWE Standards Framework

Below we describe each of the seven SWE standards.

3.1   Observations and Measurements (O&M)

The OpenGIS Observations and Measurements (O&M) Standard provides a standard model for representing and exchanging observation results. O&M provides standard constructs for accessing and exchanging observations, alleviating the need to support a wide range of sensor-specific and community-specific data formats. O&M combines the flexibility and extensibility provided by XML with an efficient means to package large amounts of data as ASCII or binary blocks.

The Observations and Measurements (O&M) Standard describes a conceptual model and XML encoding for measurements and observations. O&M establishes a high-level framework for representing observations, measurements, procedures and metadata of sensor systems and is required by the Sensor Observation Service Standard, for implementation of SWE-enabled architectures, and for general support for OGC standards compliant systems dealing in technical measurements in science and engineering.

3.2 Sensor Model Language (SensorML) [2]

The OpenGIS Sensor Model Language (SensorML) Standard provides an information model and encodings that enable discovery and tasking of Web-resident sensors and exploitation of sensor observations.[i]

The measurement of phenomena that results in an observation consists of a series of *processes* (also called *procedures*), beginning with the processes of sampling and detecting and followed perhaps by processes of data manipulation.

SensorML defines models and XML Schema for describing any process, including measurement by a sensor system, as well as post-measurement processing.

Within SensorML, everything including detectors, actuators, filters, and operators is defined as a process model. A *Process Model* defines the *inputs, outputs, parameters*, and *method* for that process, as well as a collection of metadata useful for discovery and human assistance. Because SensorML provides a functional model of the sensor system, rather than a detailed description of its hardware, each component can be included as part of one or more process chains that can either describe the lineage of the observations or provide a process for geolocating and processing the observations to higher level information.

3.3 TransducerML (TML)

The OpenGIS® Transducer Markup Language (TML) Encoding Standard is an efficient application and presentation layer communication protocol for exchanging live streaming or archived data to (i.e. control data) and/or sensor data from any sensor system. A sensor system can be one or more sensors, receivers, actuators, transmitters, and processes. A TML client can be capable of handling any TML enabled sensor system without prior knowledge of that system.

The protocol contains descriptions of both the sensor data and the sensor system itself. It is scalable, consistent, unambiguous, and usable with any sensor system incorporating any number sensors and actuators. It supports the precise spatial and temporal alignment of each data element. It also supports the registration, discovery and understanding of sensor systems and data, enabling users to ignore irrelevant data. It can adapt to highly dynamic and distributed environments in distributed net-centric operations.

The sensor system descriptions use common models and metadata and they describe the physical and semantic relationships of components, thus enabling sensor fusion.

TML was introduced into the OGC standards process in 2004 and is now part of the SWE family of standards. It complements and has been harmonized with SensorML and O&M. TML provides an encoding and a conceptual model for streaming real-time "clusters" of time-tagged and sensor-referenced observations from a sensor system. SensorML describes the system models that allow a client to interpret, geolocate, and process the streaming observations.

---

[2] SensorML got its start in earlier NASA and CEOS (Committee for Earth Observation Satellites) projects. It was brought into OGC because OGC provides a process in which this and other elements of Sensor Web Enablement could be developed in an open consensus process.

3.4 Sensor Observation Service (SOS)

The OpenGIS Sensor Observation Service Interface Standard defines an API for managing deployed sensors and retrieving sensor observation data. SOS provides access to observations from sensors and sensor systems in a standard way that is consistent for all sensor systems including remote, in-situ (e.g., water monitoring), fixed and mobile sensors (including airborne / satellite imaging). The SOS is a critical element of the SWE architecture, defining the network-centric data representations and operations for accessing and integrating observation data from sensor systems.

The SOS mediates between a client and an observation repository or near real-time sensor channel. Clients can also access SOS to obtain metadata information that describes the associated sensors, platforms, procedures and other metadata associated with observations.

Registries (also called catalogs) play an important role. The schema for each sensor platform type is available in a registry, and sensors of that type are also in registries, with all their particular information. The schema for each observable type is available in a registry, and stored collections (data sets) of such observables and live data streams of that type are also in registries. Searches on the registries might reveal, for example, all the active air pollution sensors in London. Similarly, automated methods implementing the SOS standard might be employed in an application that displays a near real-time air pollution map of the city.

3.5 Sensor Planning Service (SPS)

The OpenGIS® Sensor Planning Service (SPS) Interface Standard defines interfaces for queries that provide information about the capabilities of a sensor and how to task the sensor. The standard is designed to support queries that have the following purposes: to determine the feasibility of a sensor planning request; to submit such a request; to inquire about the status of such a request; to update or cancel such a request; and to request information about other OGC Web services that provide access to the data collected by the requested task.

An example of an environmental support system is diagrammed above in Figure 3. This system uses SPS to assist scientists and regulators in formulating collection requests targeted at water quality monitoring devices and data archives. Among other things, it allows an investigator to delineate geographic regions and time frames, and to choose quality parameters to be excluded or included.

3.6 Sensor Alert Service (SAS)

The OpenGIS® Sensor Alert Service Best Practices Paper (OGC Document 06-028r3) specifies interfaces for requesting information describing the capabilities of a Sensor Alert Service, for determining the nature of offered alerts, the protocols used, and the options to subscribe to specific alert types. The document defines an alert as a special kind of notification indicating that an event has occurred at an object of interest, which results in a condition of heightened watchfulness or preparation for action. Alerts messages always contain a time and location value. The SAS acts like a registry rather than an event notification system. That is, the SAS will not send any alerts. All actual messaging is performed by a messaging server.

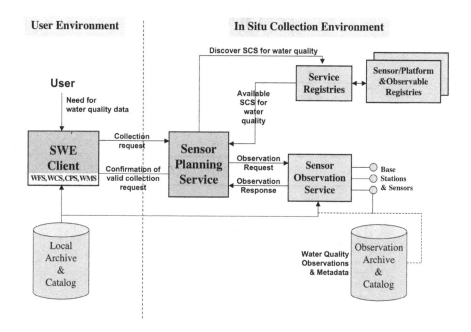

(Image courtesy of the OGC)

Fig. 3. Typical in situ Sensor Planning Service

3.7 Web Notification Service (WNS)

The OpenGIS® Web Notification Service (WNS) Best Practices Paper (OGC Document 06-095) specifies an open interface for a service by which a client may conduct asynchronous dialogues (message interchanges) with one or more other services. As services become more complex, basic request-response mechanisms need to contend with delays and failures. For example, mid-term or long-term transactions demand functions to support asynchronous communications between a user and the corresponding service, or between two services, respectively. A WNS is required to fulfill these needs within the SWE framework.

4 Sensor Web Standards Harmonization

4.1 IEEE 1451 Transducer Interfaces

An open standards framework for interoperable sensor networks needs to provide a universal way of connecting two basic interface types – transducer interfaces and application interfaces. Specifications for transducer interfaces typically mirror hardware specifications, while specifications for service interfaces mirror application requirements. The sensor interfaces and application services may need to interoperate and may need to be bridged at any of many locations in the deployment hierarchy.

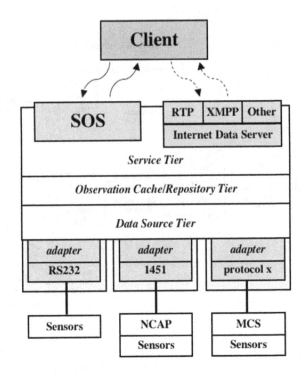

Fig. 4. IEEE-1451 in the SWE Interoperability Stack (Image courtesy of the OGC)

At the transducer interface level, a "smart" transducer includes enough descriptive information so that control software can automatically determine the transducer's operating parameters, decode the (electronic) data sheet, and issue commands to read or actuate the transducer.

To avoid the requirement to make unique smart transducers for each network on the market, transducer manufacturers have supported the development of a universally accepted transducer interface standard, the IEEE 1451 standard.

The object-based scheme used in 1451.1 makes sensors accessible to clients over a network through a Network Capable Application Processor (NCAP), and this is the point of interface to services defined in the OGC Sensor Web Enablement standards. In Figure 4, SWE services such as SOS act as clients (consumers) of IEEE-1451 NCAP services and TEDS documents, thereby enabling interactions with heterogeneous sensor systems via scalable networks of applications and services.

In addition to IEEE 1451, the SWE standards foundation also references other relevant sensor and alerting standards such as the OASIS Common Alerting Protocol (CAP), Web Services Notification (WS-N) and Asynchronous Service Access Protocol (ASAP) standards. OGC works with the groups responsible for these standards to harmonize them with the SWE standards.

4.2 Imaging Sensors

The SWE sensor model supports encoding of all the parameters necessary for characterizing complex imaging devices such as those on orbiting earth imaging platforms. ISO and OGC have cooperated to develop two ISO standards that are relevant to the SWE effort: ISO 19130 Geographic Information – Sensor and Data Model for Imagery and Gridded Data and ISO 19101-2 Geographic Information – Reference Model – Imagery (OGC Abstract Specification, Topic 7). Other related work for support of imaging sensors within the SWE context include: OpenGIS® Geography Markup Language (GML) Encoding Standard, GML Application Schema for EO Products Best Practices Paper (OGC Document 06-080r2), OpenGIS® GML in JPEG 2000 for Geographic Imagery Encoding Standard and OpenGIS GML Encoding of Discrete Coverages Best Practices Paper (OGC Document 06-188r1).

5 Current Implementation Efforts

Below are descriptions off some current SWE implementation efforts.

5.1 NASA

Adopting sensor webs as a strategic goal, the U.S. National Aeronautics and Space Administration (NASA) has funded a variety of projects to advance sensor web technology for satellites. A number of these projects have adopted the OGC's Sensor Web Enablement (SWE) suite of standards. Central to many of these efforts has been the collaboration between the NASA Jet Propulsion Lab and the NASA Goddard Space Flight Center (GFSC) using the Earth Observing 1 (EO-1) and assorted other satellites to create sensor web applications which have evolved from prototype to operational systems.

In the OWS-4 test bed activity, GSFC, Vightel Corp., and Noblis initiated a sensor web scenario to provide geographic information system (GIS)-ready sensor data and other infrastructure data to support a response to a simulated wildfire emergency. One of the satellites used in the demonstration was NASA's EO-1, which had the nearest next in-time view for the target, and so provided the real-time data used. An OWS-4 team prototyped the preliminary transformation to SWE implementation using the Open Source GeoBliki framework they developed for this purpose. Both JPL and GSFC are in the process of changing the remaining EO-1 interfaces to OGC SWE compatibility. Figure 5 shows some of the other missions that have begun to adapt portions of the standard. NASA is using the SWE standards to standardize and thus simplify sending commands to satellites.

5.2 SensorNet®

SensorNet®[3] is a vendor-neutral interoperability framework for Web-based discovery, access, control, integration, analysis, exploitation and visualization of online sensors,

---

[3] http://www.sensornet.gov

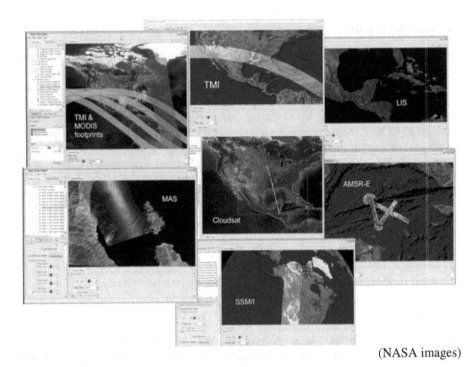

(NASA images)

Fig. 5. All of these satellite and airborne sensors are, at least some of the time, using SensorML for geolocation and other purposes. See NASA's JPL and GSFC Sensorweb/EO-1 pages, http://sensorweb.jpl.nasa.gov/ and http://eo1.gsfc.nasa.gov/ .

transducers, sensor-derived data repositories, and sensor-related processing capabilities. It is being designed and developed by the Computational Sciences and Engineering Division at Oak Ridge National Laboratory (ORNL), in collaboration with the National Oceanic and Atmospheric Administration (NOAA), the Open Geospatial Consortium (OGC), the National Institute for Standards and Technology (NIST), the Institute of Electrical and Electronics Engineers (IEEE), the Department of Defense, and numerous universities and private sector partners. The purpose of SensorNet is to provide a comprehensive nationwide system for real-time detection, identification, and assessment of chemical, biological, radiological, nuclear, and explosive hazards.

The SensorNet team is developing prototypes based on standards and best practices to network a wide variety of sensors for strategic testbeds at military installations, traffic control points, and truck weighing stations. The sensor networks will be connected by secure and redundant communication channels to local, regional and national operations centers. These network testbeds will provide a basis for interfaces to 911 centers, mass notification networks, automated predictive plume modeling applications and evacuation models.

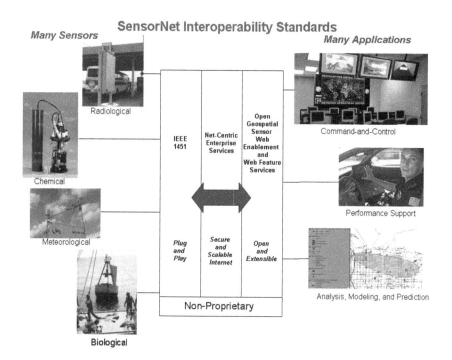

(Image courtesy of ORNL)

Fig. 6. SensorNet lays the groundwork for rapid deployment of a nationwide real-time detection system

5.3 HMA in Europe

The European Space Agency and various partner organizations in Europe are collaborating on the Heterogeneous Mission Accessibility (HMA) project. HMA's high-level goals include consolidating earth imaging and other geospatial interoperability requirements; defining interoperable protocols for cataloging, ordering, and mission planning; and addressing interoperability requirements arising from security concerns such as authorization and limiting reuse. HMA involves a number of OGC standards, including the Sensor Planning Service, which supports the feasibility analysis requirements of Spot Image optical satellite missions (Figure 7).

5.4 Northrop Grumman's PULSENet

Northrop Grumman Corp. (NGC) (http://www.northropgrumman.com) has been using the SWE standards in a major internal research and development (IRAD) project called Persistent Universal Layered Sensor Exploitation Network (PULSENet) (Figure 8). This real-world test bed's objective is to prototype a global sensor web that enables users to:

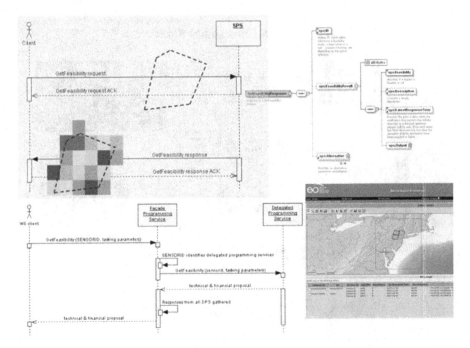

Image courtesy of ESA

Fig. 7. SPS GetFeasibility operation in a single and multiple satellite environment

- Discover sensors (secure or public) quickly, send commands to them, and access their observations in ways that meet user needs
- Obtain sensor descriptions in a standard encoding that is understandable by a user and the user's software
- Subscribe to and receive alerts when a sensor measures a particular phenomenon

In its first year, PULSENet was successfully field tested under a real-life scenario that fused data from four unattended ground sensors, two tracking cameras, 1,800 NOAA weather stations and the EO-1 satellite.

5.5 SANY Sensors Anywhere

SANY IP (http://www.sany-ip.org/) (Figure 9) is co-funded by the Information Society and Media Directorate General of the European Commission. SANY IP intends to contribute to Global Monitoring for Environment and Security (GMES, a major European space initiative), and the Global Earth Observation System of Systems (GEOSS) by developing a standard open architecture and a set of basic services for in situ sensor integration of all kinds of sensors, sensor networks, and other sensor-like services. It aims to improve the interoperability of in-situ sensors

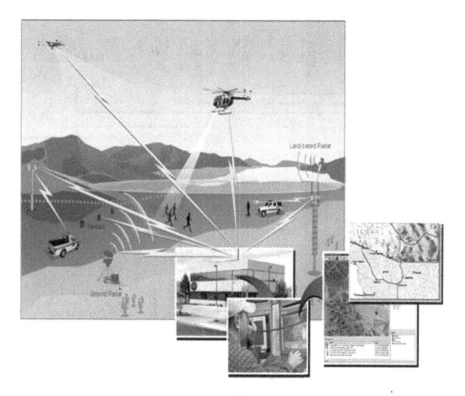

(Image courtesy of Northrop Grumman)

Fig. 8. PULSENet clients in multiple Web locations can task heterogeneous sensors and sensor systems

and sensor networks and to allow quick and cost-efficient reuse of data and services from currently incompatible sources for future environmental risk management applications. Though SANY addresses interoperability in monitoring sensor networks in general, it focuses on air quality, bathing water quality, and urban tunnel excavation monitoring.

The SANY Consortium recognizes the OGC's SWE suite of standards as one of the key technologies that can eventually lead to self-organizing, self-healing, ad-hoc networking of in situ and earth observation sensor networks. Earlier this year, SANY evaluated the capabilities of SWE services with the intention of actively contributing to further development of the SWE standard specifications. As reported at the OGC TC meeting in Ispra (December 2007), the SANY Consortium has described the common architecture used in SANY baseline applications; included architectural requirements inherited from ORCHESTRA, GEOSS, etc.; and published a road map for v1, v2, and v3 versions of the architecture.

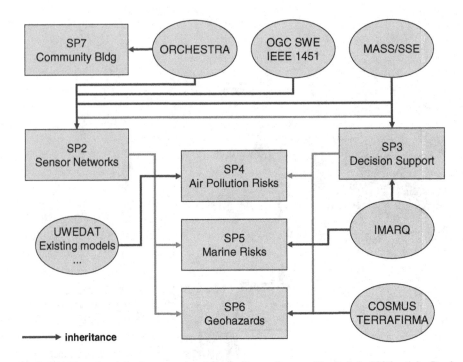

Fig. 9. SANY project inheritance and activities. (Image from an Enviroinfo 2006 article, Denis Havlik et al, "Introduction to SANY (Sensors Anywhere)Integrated Project" In: Klaus Tochtermann, Arno Scharl (eds).

5.6 52North

The German organization 52°North provides a complete set of SWE services under GPL license. This open source software is being used in a number of real-world systems, including a monitoring and control system for the Wupper River watershed in Germany and the Advanced Fire Information System (AFIS), wildfire monitoring system in South Africa.

One of several research projects using 52°North's software is the German Indonesian Tsunami Early Warning System (GITEWS) (Figure 11), a 35-million-euro project of the German aerospace agency, DLR, and the GeoForschungsZentrum Potsdam (GFZ), Germany's National Research Centre for Geosciences. GITEWS will use SWE services as a front-end for sharing Tsunami-related information among the various components of the GITEWS software itself. GITEWS uses real-time sensors, simulation models, and other data sources, all of which must be integrated into a single system. SANY, mentioned earlier, is also using 52°North's software.

5.7 Access to U.S. Hydrologic Data

The Consortium of Universities for the Advancement of Hydrologic Science Inc. (CUAHSI) is an organization founded to advance hydrologic research and education

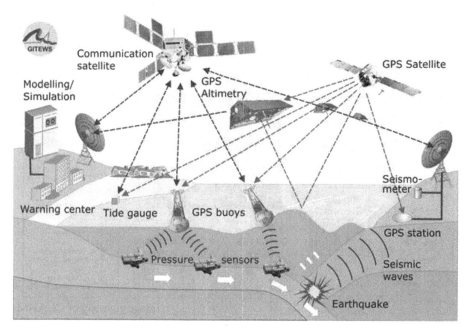

(Image courtesy of GITWS)

Fig. 10. German Indonesian Tsunami Early Warning System (GITEWS) components

by organizing and supporting university-based collaborative projects. CUAHSI represents more than 100 U.S. universities, as well as international affiliates, and is supported by the U.S. National Science Foundation. Its Hydrologic Information System (HIS) project involves several research universities and the San Diego Supercomputer Center as the technology partner. For three years, the CUAHSI HIS team has been researching, prototyping, and implementing Web services for discovering and accessing different hydrologic data sources, and developing online and desktop applications for managing and exploring hydrologic time series and other hydrologic data.

The core of the HIS design is a collection of WaterOneFlow SOAP services for uniform access to heterogeneous repositories of hydrologic observation data. (SOAP is a protocol for exchanging XML-based messages over computer networks. SOAP forms the foundation layer of the Web services stack, providing a basic messaging framework that more abstract layers can build on.) The services follow a common XML messaging schema named CUAHSI WaterML, which includes constructs for transmitting observation values and time series, as well as observation metadata including information about sites, variables, and networks. At the time of writing, the services provide access to many federal data repositories (at USGS, EPA, USDA, NCDC), state and local data collections, as well as to data collected in the course of academic projects. The HIS Server, which supports publication of hydrologic observations data services, is deployed at 11 NSF-supported hydrologic observatories across the country and a number of other university sites (Figure 11).

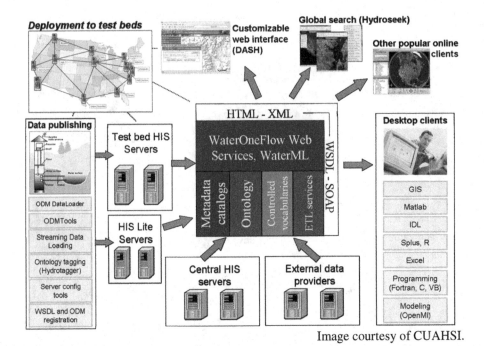

Image courtesy of CUAHSI.

Fig. 11. WaterOneFlow Web services will provide a standard mechanism for flow of hydrologic data between hydrologic data servers (databases) and users

The WaterML specification (Figure 11) is available as an OGC discussion paper (document 07-041). The CUAHSI HIS team is working with OGC to harmonize it with OGC standards such as GML and the OGC Observations and Measurements specification to make the next version of CUAHSI Web services OGC-compliant. The project's web site is www.cuahsi.org/his. The project is supported by NSF award EAR-0622374, which is gratefully acknowledged.

6 Conclusion

OGC's SWE standards have the potential to become key parts of an integrated global framework for discovering and interacting with Web-accessible sensors and for assembling and utilizing sensor networks on the Web. OGC invites additional participation in the consensus process and also invites technical queries related to new implementations of the emerging standards.

Linking Geosensor Network Data and Ontologies to Support Transportation Modeling

Kathleen Stewart Hornsby[1,2] and Kraig King [2]

[1] National Center for Geographic Information and Analysis
[2] Department of Spatial Information Science and Engineering
University of Maine
Orono, ME 04469-5711
{kathleen,kking}@spatial.maine.edu

Abstract. This work discusses the supporting role of ontologies for geosensor network data focusing on methods for linking geosensor network datasets with ontologies. A major benefit gained from this kind of linking is the augmentation of databases with generalization or specialization relations from an ontology. We present methods for linking based on a transportation application where data on vehicle position is collected from sensors deployed along a road network and stored in a geosensor database. A mechanism for linking, implemented as a tab widget within the Protégé ontology editor, is introduced. This widget associates data values from the geosensor database with classes in the ontology and returns linked terms (i.e., matches) to the user as a set of equivalence relations.

Keywords: transportation ontologies; geosensor network data; moving object databases; linking ontologies.

1 Introduction

Where data collected from geosensor networks are capable of giving users a more-or-less continual record of measured values with respect to geospatial phenomena, these data streams by themselves do not necessarily give a context for the data that includes semantics. Combining such data streams with ontologies, however, provides a foundation for deriving a real-time understanding of dynamic geospatial domains that incorporate semantics. Ontologies are a structured way for describing the characteristics and relationships of objects found in the world around us. For example, a network of sensors placed along a roadway measures the positions of vehicles at fixed reference points over time. These measurements in combination with supporting ontologies provide a basis for reasoning about, for example, the type of vehicles traveling on the road network (e.g., is the vehicle traveling behind my automobile a police car?) as well as semantic similarities between them (e.g., both the police car and ambulance traveling ahead of me are a type of emergency road vehicle). The details of the objects and their associated positions are derived from data streams and stored in databases, while generalizations or refinements of the moving objects are supplied by ontologies.

S. Nittel, A. Labrinidis, and A. Stefanidis (Eds.): GSN 2006, LNCS 4540, pp. 191–209, 2008.

Ontologies can play a role in the integration and combination of data streams from different sources (e.g., images from traffic cams combined with data from fixed-location sensors along a road) by providing the foundation for discovering the similarities between the sensor data sources. The semantic descriptions supplied by ontologies enable computers to process these geosensor data streams, such that their data can be extended and reused. Ontologies contribute to this *interoperability* by providing additional perspectives that are intelligible by both computers and humans. This feature makes possible machine learning algorithms that support linking mechanisms for geosensor data.

In this work we consider methods for combining data from geosensor networks by linking the databases storing sensor data with ontologies that model the objects moving within a transportation domain. Existing applications such as MAPONTO and VisAVis map database schemas with elements of the ontology. However, more information may be desired by means of linking the instance-level data stored within the geosensor database to the class and attribute names in a related ontology. Users can leverage this integration of geosensor data streams to aid in their understanding of the semantics of the objects traveling on the road network around them. For example, a cautious driver may want to know the kinds of emergency vehicles surrounding their vehicle. This research introduces a linking mechanism, implemented as a tab widget within the Protégé ontology editor [1], that will be used to intuitively connect geosensor network data with an ontology to increase or decrease information granularity.

The remainder of this chapter is structured as follows: a discussion of related research on geospatial ontologies as well as streaming data collected within a geosensor network is presented and a framework for collecting geospatial positional data for moving objects on a road network is introduced. In addition, a database structure for storing the positional data is proposed. Based on the available classes of objects originating from existing transportation ontologies, an ontology of transportation devices is introduced. Finally, a formal framework and rationale for linking geosensor network data with the *TransportationDevice* ontology is presented. Conclusions and plans for future work are discussed in the final section of the chapter.

2 Related Work

The collection, organization, analysis and delivery of geospatial information from distributed sensor networks is an active research area (see, for example, [2]). Tracking patterns of moving objects (i.e., vehicles) is a subfield of geosensor research, with one focus relating to the modeling of moving objects via sequences of location-time pairs that form *trajectories* [3, 4, 5, 6, 7, 8]. Trajectories are fundamental for tracing past and current object movement as well as predicting future motion plans [9]. Additional research has focused on hybrid representations for modeling moving objects, such as nonmaterialized trajectories, in an effort to overcome location imprecision due to sensor error [10].

Data models that support the modeling of moving objects in a geosensor network have been studied both in a geographic information science context as well as in computer science. Research has been conducted on moving object databases [11, 12, 13, 14] where some of the topics include querying moving object databases [15, 16], indexing [17], modeling moving objects over multiple granularities [18] and modeling moving objects on a road network using geographic data technology maps [19]. The real-time characteristic of moving objects introduces challenges for managing sensor data, such as determining update intervals, and dealing with imprecision and uncertainty regarding an object's location. Techniques such as dead-reckoning [11], point location management and trajectory location management [20] are some of the proposed solutions.

Research into developing geospatial ontologies has explored the prime geospatial categories and concepts that underlie such ontologies, highlighting, for example, basic geographic features such as mountains, rivers, and lakes [21, 22]. These ontologies are especially useful for supporting geographic information integration in a seamless and flexible way [23]. Establishing links between ontologies and database content is still a relatively new area of investigation. Delivering content for semantic web applications is encouraging further research, however, and automated methods for mapping between the database schema and ontologies are being explored [24, 25].

The management and integration of databases and ontologies is an additional topic of interest for researchers. The diversity of computational resources that provide geosensor data, (e.g., sensors, satellites, embedded processors, or GPS) must be properly managed to create a fully collaborative system that offers transportation solutions such as autonomous real-time driving, routing and navigation [9]. Our work contributes to designing next-generation transportation information architectures by proposing a method for relating these geosensor databases and ontologies.

Ongoing research relating to the supporting role of ontologies for geosensor data has been investigated, with a focus on methods for connecting geosensor network data with ontologies that are modeled using the Protégé ontology editor [1]. Various terms have been used to derive connecting either multiple ontologies or ontologies and databases [26]. These terms and associated definitions include: *aligning*: two or more ontologies are brought into mutual agreement so that they appear consistent and coherent; *combining* where two or more ontologies that have similar elements (e.g., classes, attributes) are used in such a way that they act like a single unit; *mapping* is the relating of similar elements from different sources with an equivalence relation such that they appear to be integrated virtually; and *merging* where creating a new ontology is created from two or more ontologies that contain overlapping elements.

Each of these terms conveys a semantic meaning that pertains to the relation between classes or elements of different ontologies. The first of these terms, *aligning* doesn't necessarily specify if a new ontology is created, nor if the original ones persist. *Combining*, on the other hand, recognizes the similarities between ontologies and specifies they are treated as a single unit, but again, it is not known if this is a new ontology. In contrast, *mapping* specifies that the resulting ontology appears to be integrated, perhaps implying that the parents persist while a new 'virtual ontology' is created. Another relation, *merging,* specifies that the parent ontologies are integrated to form a new, independent ontology.

In this research, we focus on connecting geosensor databases and ontologies, rather than pairs of ontologies. The term *linking* will be used to describe this process. *Linking* is similar to *mapping,* in that elements from independent databases and ontologies will be matched in order to determine what equivalence relations exist, enabling information to be shared between them. During the process of integration, the structure and elements within each source remain unchanged. We present a tool, in the form of a tab widget implemented in Protégé, which allows a user to link the complex data from a geosensor network and stored in a database with an ontology such that the individual structure and content of each one still persists. By applying this tool, a list of the equivalence relations that hold between the database and ontology is automatically generated by the system.

3 A Framework for Collecting Dynamic Geosensor Network Data within a Transportation Domain

For this work, a fixed-length linear referencing model is used in conjunction with a point-location approach to represent traffic movement data. A *fixed-length referencing system* holds spatial units constant by dividing the route into segments of uniform length [27]. *Point-location management* is a straightforward approach to modeling the location of moving objects, where a location-time pair is generated periodically for each object and then stored in a database for future analysis [20].

The Moving Objects on Dynamic Transportation Networks (MODTN) model [28] can be expanded to create such a geosensor framework for capturing vehicle movement in a transportation network. Each object moving through the network would be equipped with its own portable computing device that is interconnected with the network through some medium such as a wireless interface. Furthermore, the network itself has a number of sensors distributed along the route such that they can measure the position (and possibly distance covered by) the moving object. Whenever an object transfers from one sensor to another, a series of location updates are triggered.

The network of sensors is distributed along a roadway so that each sensor observes a specific number of fixed reference positions p. These reference positions are discretely numbered from $+1$ to $+\infty$ along a segment of highway. At time t, only one object can be at a specific reference position for any given lane (e.g., one object cannot be on top of another object). To reduce the degree of parallax introduced by line-of-sight techniques, it is assumed that pairs of sensors are placed on both sides of the road such that the first detects traffic in one direction and the second detects traffic in the opposite direction (Figure 1). We also assume that a single identifier references both sensors (i.e., a pair of sensors $\{s_1, s_{1a}\}$ is represented by the id s_1). Traffic lanes ℓ are numbered sequentially from 1 to ∞. Thus, a four-lane highway may be represented by the set $\{\ell_1, \ell_2, \ell_3, \ell_4\}$.

Data collection begins when movement is detected. A timestamp for each sensor reading is stored in a general date/time format such as *mm/dd/yyyy hh:mm:ss*. From the initial point of movement, a series of sensor readings r are collected at a fixed time t from one another. Thus, if a moving object triggers sensor s_1, a set of n readings are collected $\{r_1 t_1, r_2 t_2, r_3 t_3 ... r_n t_n\}$. The positions of each individual object within the range of the sensor (relative to a reference position) are then stored for each reading.

Each moving object is assigned a unique identifier, for example, by a mechanism such as the broadcast of a vehicle identification number. The linear extent of the object (*length*) and its midpoint (*position*) are stored as well. *Length* is calculated as a function of the reference position interval within the sensor network (e.g., an object length of 1.5 is equivalent to 1.5 reference position intervals).

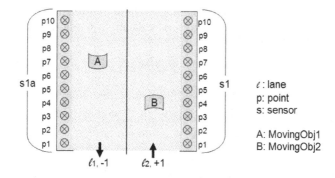

Fig. 1. Infrastructure for collecting moving object data

The location of a moving object is encoded by obtaining the corresponding reference position in the sensor network that is closest to the midpoint of that object. If necessary, a second relation can be constructed that explicitly defines the location as a set of coordinates (e.g., GPS or lat-long) for each of these positions. Next, the lane in which an object is moving must be stored. Each lane is assigned an *ID* and *direction* that is based on one of two possible values: flow either follows the sequencing of sensors (ℓ_2,+1) or is against the sequencing of sensors (ℓ_1,-1) (Figure 1).

3.1 A Database Representation for Storing Geosensor Data

In order to develop a basic framework for storing geosensor data relating to a transportation network, special consideration must be given to the collection of the object data given that it is characterized by discrete positional changes over time. Conventional database management systems assume data remains constant unless it is modified. In contrast, geosensor data streams use continuous location data that are sampled such that they meet the desired granularity requirements. This characteristic demands frequent database updates to ensure that inaccurate and outdated data is not stored [29]. The MODTN framework [28] is leveraged to manage these database updates with an ID-Triggered Locations Update (ITLU) schema. Whereas in the MODTN framework groups of sensors are positioned at route intersections, we assume sensors are distributed uniformly along the road network. Whenever an object transfers from one sensed position to another, a location update will be triggered to store the object's measured position within the geosensor database.

Additional details of the object and its associated movement are also stored within the geosensor database. A data model describing the structure for capturing details of a mobile object in a moving objects databases (MODs) has been proposed in [11]. Along with the object's unique ID, attributes such as route, start location, start time, direction, speed and uncertainty are stored. In a similar fashion, the database representation used in this work to store the geosensor positional data depends on a relation *SensorDat* with attributes, *objectID*, *sensorID*, *laneID*, *position*, and *time*. This relation stores location readings within the sensor from the network. Details of the spatial representation of the moving objects, on the other hand, are stored in relation *ObjData* with attributes, *objectId*, *objType* and *length*. The attribute *objType* corresponds to object classes in the moving objects ontology. These two relations, *SensorDat* and *ObjData*, provide the foundation for a geosensor database that stores motion data captured within the sensor network (Figure 2).

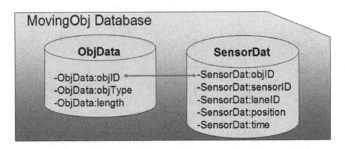

Fig. 2. Database structure for storing moving object data

3.2 Ontologies That Model Entities in a Transportation Domain

There are many contexts from which a geospatial domain can be modeled. Each of these contexts contributes to the development of an ontology that describes the constituents of the domain. For a transportation domain, one possible context is based on the entities that are commonly encountered on a transportation network. These entities either move on the actual transportation network (e.g., ambulances or buses), or move in such a way that they interact with the other entities traveling on the road network, (e.g., cyclists). Each kind of moving entity can be modeled as a class. These classes, their attributes, and the relations among the classes serve as a foundation for building a transportation ontology composed of moving entities.

In this work, an ontology based primarily on the classes of entities that are commonly encountered moving on a road network is established. It is assumed that the moving entities are land vehicles that travel about on a transportation network composed of roadways. Train-, airplane-, and waterway-related classes will not be considered because they are outside of this scope. These moving entities are derived from the mid-level ontology of land vehicles that are extracted from the Suggested Upper Merged Ontology (SUMO) knowledge base developed for the IEEE (http://sigma.ontologyportal.org). The classes, related by *is_a* relations, form an ontology of transporation devices (Figure 3). The *TransportationDevice* ontology has three primary superclasses: *RoadVehicle*, *Cycle*, and *RailVehicle*. The class *RoadVehicle* has subclasses *EmergencyRoadVehicle*, *Truck*, *Automobile* and *MotorCycle*. *Automobile* is further subdivided into subclasses *TaxiCab*, *Bus,* and *PassengerCar*. It should be noted that *Bus* is modeled as a subclass of *Automobile* based on the fact that SUMO defines a bus to be, "An automobile designed to carry at least a dozen people… …that can transport large numbers of passengers (i.e., dozens) at one time". The class *EmergencyRoadVehicle* has specializations, *FireEngine* and *Ambulance*. Within the SUMO framework, *PoliceCar* is considered to be a subsuming mapping of the class *Automobile*. However, SUMO defines an *EmergencyRoadVehicle* to be: *"a RoadVehicle designed for special use in emergencies."* Therefore, in this work *PoliceCar* is modeled as a subclass of *EmergencyRoadVehicle*.

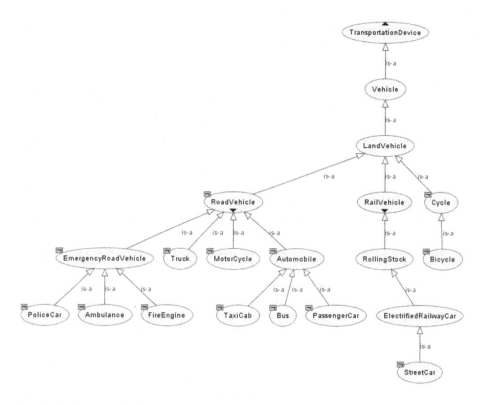

Fig. 3. A *TransportationDevice* ontology from the SUMO knowledge base (http://sigma.ontol-ogyportal.com)

Although most classes of moving objects on the road network are subsumed by the class *RoadVehicle*, we extend the ontology to also include other subclasses of SUMO class *LandVehicle* such as *Cycle* and *RailVehicle*. This is due to the fact that bicycles and street cars are two types of moving objects that also interact with other land vehicles on the road network. For example, a streetcar interacts with other vehicles by moving on tracks that are laid into the street.

4 Linking Geosensor Network Data and Ontologies

To maximize the benefit of geosensor data streams for a transportation system, mechanisms need to be developed for combining and extending this data. Ontologies provide structure for the geosensor data and allow both humans and machines to perform reasoning and make inferences that are either more generalized or more specialized as necessary. A number of tools have been created to link ontologies by exploiting the similarities between them. For example, applications such as CRAVE [30], OLA [31], oMAP [32], and PROMPT [33], each provide a semi-automated system for ontology alignment. Others, such as MAPONTO [34] and VisAVis [24], provide a mechanism for connecting relational database schemas with ontologies. The

MAPONTO tool assists users in discovering semantic relationships that exist between a database schema and a target ontology. The resulting output is a set of rules that express these semantic mappings. Similarly, VisAVis compares the database schema with an ontology by employing a graphical interface that is developed within the Protégé ontology editor. VisAVis outputs the resulting relations within a new ontology by augmenting the initial ontology with SQL references to the mapped database.

Our research builds upon the technique of mapping databases to ontologies. It looks beyond the database schema to link the instance-level data contained within the geosensor network dataset to the classes and attributes of a related ontology. The linking of geosensor databases and ontologies provides a mechanism for combining independent sources so that information can be shared between them. Thus, in contrast to other methods such as merging, the linking process minimizes storage requirements because a new database or ontology need not be created. Relations are dynamically generated between the database and ontology from within the ontology editor Protégé (http://sigma.ontologyportal.org, [35]). The linking mechanism we developed locates all possible database matches by searching the classes and attributes of a related ontology and returns these matches as *equivalence relations*.

The linking process generates a list of equivalence relations by employing a series of steps that involve *specification, parsing, matching,* and *granularity control.* First, the user must specify the source geosensor database, the target ontology and the key elements of each that are to be connected. Once these entities are identified, the linking application iteratively selects each database value and parses the ontology structure for a potential match. If a match is found, an equivalence relation is generated. Once a list of equivalence relations has been created, the user has the ability to choose the desired granularity of the results, thereby choosing a higher-level and more abstract understanding of their domain, or a more refined view depending on their needs. The following sections will discuss in greater detail, a system architecture, which employs each of these linking steps.

4.1 Specification: Identifying Key Elements for Linking

The first step in this linking process is the identification of key elements from both the geosensor database and ontology that will be used for connecting these entities. These key elements become possible attachment points for linking. For the geosensor database, key elements may be specific *tuple values* or components of the database schema such as *attribute name, attribute datatype* and additional *attribute metadata.* Attribute metadata refers to supplementary data that further describes a relation (e.g., attribute definitions, search keys, etc.). For an ontology, the key elements that serve as a basis for linking are *class name, attribute name* and possibly *instances of classes* if they are defined.

Existing systems such as MAPONTO and VisAVis map a database schema with elements of the ontology via lexical matches. In this work, we focus on an alternative method of linking, where individual tuple values from the geosensor database are linked with the classes, attributes and instances found within a related ontology. This is a type of *deep linking* where the attachment points are the instance data contained within the geosensor database rather than its schema. An example of this deep linking process is provided later in section 4.3.

In this linking application, two possible options for filtering source data from the geosensor database are available. The first option transpires when the user does not specify which attribute names they are most interested in linking. If no attribute names are specified, the linking application assumes that all attributes are to be processed for possible matches. The second option leads to a reduction of the source data per the user's choice of attribute names. This is achieved by applying the relational projection operator, π, via an internal application query. This reduces the search space so that only values of interest from the geosensor database are parsed for semantic equivalence. For example, the *ObjData* relation discussed in section 3.1 contains three attributes: *objID*, *objType* and *length*. The linking application, by default, will process instances from all three attributes for corresponding ontology matches. However, a user has the ability to declare that matches only be found for a specific attribute (e.g., *objType*). For the relation *ObjData*, this reduces the number of database instances to be processed by approximately sixty-six percent. Thus, the ability to specify which attributes are processed for matches provides the user with a mechanism to filter the results and reduce the processing time of the linking application.

4.2 Parsing: Searching for Linking Elements

To facilitate linking, an algorithm has been designed to iterate through the ontology structure looking for linking elements (e.g., class names, attribute names, and instance values). We leverage the parsing conventions offered by the Protégé-OWL advanced programming interface (i.e., API). The Protégé-OWL API is an open-source Java library that allows programmers to create and manipulate source files that have been developed using the Web Ontology Language (OWL) and Resource Description Framework (RDF). Within the API, class names that are stored within the target ontology are identified by making calls to the *getSubclasses()* routine. This routine returns a collection of descendent classes for any given class.

A simple recursive function is used to traverse the class hierarchy of the ontology (Figure 4). The *getSubclasses()* command has a Boolean parameter, that if set to true, will not only return the direct subclasses, but also their children, grandchildren, etc. In order to facilitate the individual examination of each subclass within the ontology, this flag is set to false and traversal of the ontology is controlled by the recursive function that has called it. As soon as a collection of descendant subclass names has been stored, additional logic can be employed to make use of this data.

This recursive function can be modified to discover additional characteristics of each class such as attribute names and instances. For example, the function *getTemplateSlots()* can be used to retrieve a collection of attributes for a given class (attributes are referred to as *slots* within Protégé). Additionally, a collection of available instances for each class can be returned with the *getInstances()* routine. Both of these functions can be included in the body of the *parseClasses()* function to retrieve additional information about each class.

```
private static void parseClasses(RDFSClass cls) {
 cls.getName();
 for (Iterator i = cls.getSubclasses(false).iterator(); i.hasNext();)
 {
     //Store the retrieved subclass names
     RDFSClass subclass = (RDFSClass) i.next();
     //Process the retrieved subclass names and-or data

         .

         .

     //Retrieve additional subclasses
     parseClasses(subclass);
 }
}
```

Fig. 4. Recursive function to parse an ontology by using Protégé-OWL API calls

4.3 Matching: Finding Equivalence Relations between a Database and Ontology

During the parsing process, the linking application must determine if the identified elements from the ontology and geosensor database share some similarity, such that an equivalence relation is defined. The matching process begins with the database tuple value serving as the *pattern* (i.e., character sequence) for which searched. Possible *candidates* for a match will be sought from the class names, attribute values or instances of the related ontology. For example, consider the relation *ObjData* that contains the attribute instance *taxicab* (Figure 5). The linking mechanism locates all occurrences of the instance *taxicab* by searching every class name and attribute name of the related TransporationDevice ontology. As a result, P-C pairs that include, for example, (*taxicab | transportationdevice*), (*taxicab | vehicle*), and (*taxicab | landvehicle*) would be generated during this comparison. Of these pairs, the only equivalence relation to be defined would be (*taxicab | taxicab*).

To find an equivalence relation, each tuple value of the database is compared to all available elements of the ontology. Therefore, the total number of iterations of this algorithm will be $dN*oN$, where dN is the number of tuple patterns and oN is the number of ontology candidates. The algorithm processes each pattern and candidate as an independent pair to determine if the pair is a valid attachment point. The entire set of pattern-candidate (P-C) pairs is formally represented as: $\{P_1C_1, P_1C_2... P_1C_{oN}, P_2C_1, P_2C_{oN}... P_{dN}C_{oN}\}$.

The processing of these P-C pairs for possible matches is accomplished with a standard semantic comparison. Lexical pattern matching is applied to each potential pair of geosensor database and ontology terms. To simplify the matching process, each possible pattern and match pair is normalized by ensuring that their constituent

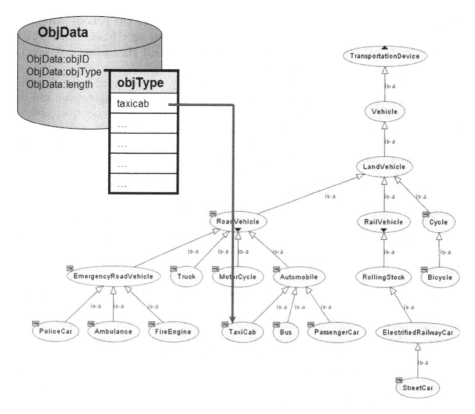

Fig. 5. Linking a geosensor database and the *TransportationDevice* ontology

characters are all lower case. Making the linking algorithm case insensitive ensures that the maximum number of possible matches is returned by the application.

The application user is provided with four pattern matching options that are used to influence the linking mechanism between the geosensor database and ontology. These options are: *direct, prefix, suffix* and *inclusion*. The first linking option, *direct*, utilizes a straightforward one-to-one comparison mechanism that is analogous to a logical equality operator. An example of a successful direct comparison would be an equivalence relation between the database value *ambulance* and the ontology class name *ambulance*.

The remaining three matching options employ a wildcard character to search for a range of candidate values. Similar to other programming languages, the character * can be substituted for a string of zero or more characters within the candidate. When determining the semantics of these matching options, the position of the pattern within the candidate should be considered. For instance, the prefix option would produce an equivalence relation for any expression of the form [pattern] = [pattern]*. Thus, a prefix based match would be generated between the tuple instance **police** and class name **policecar**. In contrast, the suffix option produces equivalence relations for expressions of the form [pattern] = *[pattern]. The tuple instance **cyclist** and class name **bicyclist** would satisfy the requirements of a suffix-based match. Alternatively, inclusion is a concatenation of both the prefix and suffix matching options. An inclusive match searches for expressions that comply with the specification [pattern] = *[pattern]*. A relation between the tuple instance **road** and class name *emergencyroadvehicle* illustrates an inclusive match.

The cardinality of these matches is represented by one of three forms. Several matches may be generated for a single source entity, that is, a tuple value can be linked to multiple ontology elements (i.e., 1:n). The converse is also true; several source entities can be linked to a single target entity (i.e., n:1). However, it is not required that every tuple within the source database form an equivalence relation with some element of the related ontology (i.e., 1:0).

In future work, the definition of an equivalence relation may become more relaxed. Although we have assumed that a strict semantic match is required, we anticipate that this constraint will be modified to include patterns that support inexact or similar matches as well as spatial and temporal matches between the geosensor database and an ontology. For example, a set of equivalence relations may be derived from coordinate data and an ontology containing geometric shapes and patterns.

5 A Tab Widget for Linking a Geosensor Database and Ontology

The framework that has been discussed for linking a geosensor database with an ontology can be drawn upon to extend existing ontology editors. Perhaps the most significant advantage of extending an ontology editor such as Protégé is the ability to leverage existing functions that are capable of manipulating ontology source files. Additional benefits include a standardized development platform, and a programming interface that enforces a basic design rationale.

The mechanism used for linking within the Protégé editor, is a plug-in that is referred to as a *tab widget*. A tab widget is essentially an extension of the core Protégé ontology editor. It is a simple user interface that appears in the main window of the Protégé editor as a clickable tab. The new tab for the linking application is positioned alongside other system tabs such as *Classes*, *Properties*, *Individuals* and *Forms* (Figure 6).

The standard tab widgets within Protégé, utilize a *frame* to arrange and display data about the ontology and its components. Each frame is composed of a widget body that contains a series of *panes* that contain various components used for sharing information and interfacing with the application user (Figure 6). These components can include list boxes, buttons, tables, text fields and other objects commonly found on standard graphical user interfaces. In the next section, we illustrate the look and functionality of the linking widget with an example from the transportation domain that was discussed earlier.

5.1 Structure of the Geosensor Database-Ontology Linking Tab

The tab widget that has been developed for the linking application consists of a single pane that provides several objects with which the user must interact. Utilizing a tab widget is beneficial because Protégé provides an intuitive mechanism for viewing and parsing the structure of the target ontology. Without these advantages, the OWL-RDF

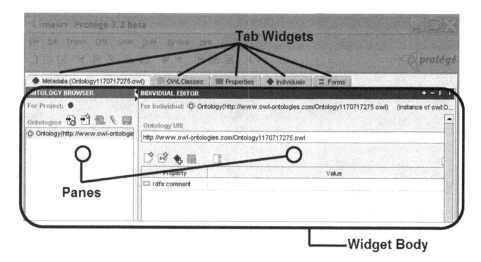

Fig. 6. Features of a Protégé tab widget

definitions of the ontology are cryptic for users to visualize and cumbersome to parse with programming code.

To use the linking tool, the Protégé ontology editor must first be launched. Within the ontology editor, the user selects and imports the desired ontology, for example, the SUMO *TransportationDevice* ontology. Once the ontology has been opened within Protégé, the user can view the associated structures using the standard tab widgets provided by the ontology editor. For example, to inspect the class hierarchy, the tab *OWL classes* should be selected.

Before the linking tab widget can be selected, the source files must first be copied to the default plug-ins folder of Protégé. The tab widget is loaded within the Protégé GUI by choosing *OWL* from the main menu and selecting *Preferences* from the list of available choices. The *Tabs* option will then provide a checkbox for enabling the linking tab widget. Once the widget becomes visible, the user will see a number of objects that are used to specify linking parameters. Proper navigation of these parameters is facilitated by the organization and layout of the graphical user interface (Figure 7).

5.2 Using the Database-Ontology Linking Tab

The core components of the linking interface are a set of objects that specify the database, ontology, matching and results criteria. The user must first identify the data source name for the geosensor database containing the moving object data, (it is assumed that an ODBC connection to this database has already been created). A subroutine within the linking widget verifies the validity of the database connection and then populates a list box object with the name of each relation (e.g., *ObjData* and *SensorDat*). This list-box is used to indicate which relation contains the data that is to be extended via the linking tool. Once the desired relation is selected, a second list-box is

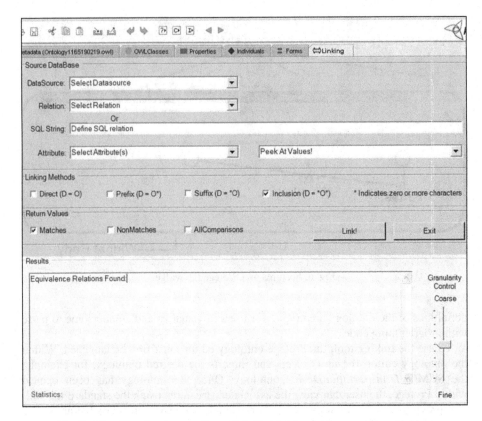

Fig. 7. Tab widget implementation within Protégé for database-ontology linking

populated with the attributes available for that relation. When one or more of these attributes are selected, the instance values can be previewed by scrolling through a third list-box. This enables the user to verify the desired data has been selected prior to running the linking tool. For example, if the *ObjType* attribute is selected, this list-box would display instance values such as *Ambulance*, *Police*, *Taxi* and *Cyclist*.

The user can proceed to choose the matching options after the instance-level values are selected. These options control the logic used for matching database instances and ontology class names. Future versions will also provide control over which additional ontology elements are linked (e.g., attribute names and available instances). One of four linking methods can be specified within the interface. These methods are *direct*, *prefix*, *suffix* and *inclusion* as discussed in section 4.3. For this example, we will elect inclusion, where each database pattern is matched with corresponding ontology elements of the form **pattern**.

The decision to return *matches* (i.e., equivalence relations), *non-matches* or all comparisons (both matches and non-matches) is another feature of the linking interface. It is anticipated that the *matches* option would be used most frequently as this option will extend the geosensor datastream knowledge. However, the *non-matches* option may also be useful in situations where one wants to create a list of all database terms that do not have a corresponding equivalence relation within the target ontology. Such a comparison may be performed, for example, by a domain expert that is looking to extend the knowledge contained within the *TransportationDevice* ontology. In rare cases, one may want to view the comprehensive set of database instances and any additional corresponding equivalence relations that are found. This list could be analyzed to perform further data analysis between the matching and non-matching terms.

Once the required inputs have been specified, the output generated by the linking application is displayed to the user within the *results* window. Each equivalence relation is shown in the form *<dbase> :: <ontology>*. If the user requests to see non-matches, only the database value is returned. Examples of equivalence relations generated from the linking of the *ObjData* relation and the *TransportationDevice* ontology would be {*ambulance :: ambulance*}, {*police :: policecar*}, and {*taxi :: taxicab*}.

In addition to displaying the equivalence relations, a statistical analysis of the linking results is provided as well. The statistics function simply compares the number of database values processed to the number of database entries that had at least one corresponding ontology match. It should be noted that this statistical analysis is influenced by the type of results that are returned. The statistical calculation may be based upon the number of matches returned, the number of non-matches returned or the number of matches *and* non-matches (always 100%). This value provides the user with a simple way to quantify the success of the linking algorithm.

A more complex linking scenario may require a combination of both the *ObjData* and *SensorDat* relations. To support this, an optional SQL query is used to specify any valid relation or composition of relations instead of selecting one from the first list box. For instance, a query is employed to determine *what types of vehicles are located between sensor positions 234 and 468?* This query requires positional data from the *SensorDat* relation and the type of object from the *ObjData* relation. It is expressed with the following statements:

```
SELECT objType, position
FROM o as objData, s as sensorDat
WHERE o.objid = s.objid
AND s.position >= 234
AND s.position <= 468;
```

In addition to displaying the matching terms of a geosensor database and an ontology, a feature has been added that allows a user to exploit the ontology further by coarsening or refining the desired granularity of the equivalence relations. The *search depth slider* is the mechanism used to augment the linking results. Increasing the search depth modifies the equivalence relations by returning ontology classes that are more specialized by finding subclasses. Conversely, decreasing the search depth returns equivalence relations that are more generalized by locating the subsuming classes. For example, if the equivalence relation {*policecar :: policecar*} is augmented through generalization, the linking application returns {*policecar :: emergencyroadvehicle*} where *emergencyroadvehicle* is coarser than *policecar* (Figure 8). If this equivalence relation is generalized further, the resulting output becomes { *emergencyroadvehicle :: roadvehicle*}. In addition to the original equivalence relations, the new generalized and specialized elements of the ontology are displayed to the user.

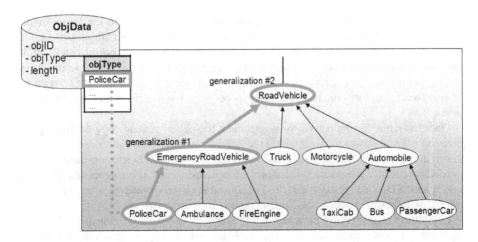

Fig. 8. Generalization of the ontology class policeCar

The first application of the slider reduces the set of matching classes from three (i.e., *ambulance, policecar, taxicab*) to two (i.e., *emergencyroadvehicle, automobile*). Further generalization reveals that all three original classes are subsumed by the class *roadvehicle*. Such generalizations would provide answers to user questions such as, *are there any emergency road vehicles between position 234 and 468?* The conventional geosensor datastream information stored within the *MovingObj* database would not be able to answer this question because the datastream by itself does not have the detail necessary for semantic reasoning. However, by linking the MovingObj database with the *TransportationDevice* ontology, and then augmenting the results by generalizing the data, such semantic descriptions and conclusions can be formed.

6 Conclusions and Future Work

This work has presented a method for linking geosensor databases with ontologies. To support this research, a geosensor data collection framework must first extract positional data for objects moving within a transportation network. The streaming positional data collected by the sensors is stored within a geospatial database. Details of the spatial representation as well as the sensed location of the moving object are collected. However, these data-streams by themselves lack the detail necessary for semantic reasoning about characteristics that are specific to each object. Ontologies such as the SUMO classification of transportation devices are leveraged to extend geosensor data streams by supplying semantic descriptions that become the foundation for discovering similarities between moving objects.

The focus of this work is the development of a mechanism for linking the geosensor database with an ontology. This mechanism is created as a tab widget within the ontology editor Protégé. The widget returns associations by locating matches between the instance data contained within the geosensor database and the class names from the related transportation device ontology. Matches are returned as a set of equivalence relations that are augmented further by employing the specialization and generalization knowledge of the ontology to provide additional perspectives of the geosensor data. Non-matches as well as all comparisons (i.e., both matches and non-matches) are also returned for the user. These perspectives provide additional details for geosensor data which are necessary for semantic reasoning and inference.

This linking mechanism will assist next-generation information systems in understanding, modeling and indexing moving objects. To further support such systems, future research will focus on spatial pattern matching where, for instance, an ontology of geometric representations can be used to augment a positional database. Such a system would facilitate the specialization and generalization of geometric shapes to increase or decrease the complexity of their spatial representation. In addition, methods for linking additional elements of a database schema with ontology class names, attribute names, and instance values are also being investigated.

Acknowledgments

Kathleen Stewart Hornsby's research is supported by a grant from the National Geospatial-Intelligence Agency HM1582-05-1-2039.

References

[1] The Protégé Ontology Editor and Knowledge Base Framework,
    http://protege.stanford.edu/
[2] Stefanidis, A., Nittel, S. (eds.): GeoSensor Networks, p. 296. CRC Press, Boca Raton (2005)
[3] Pfoser, D., Jensen, C., Theodoridis, Y.: Novel approaches in query processing for moving object trajectories. In: VLDB, pp. 395–406 (2000)
[4] Wolfson, O., Chamberlain, S., Kalpakis, K., Yesha, Y.: Modeling moving objects for location based services. Infrastructure for Mobile and Wireless Systems 2001, 46–58 (2001)
[5] Stefanidis, A., Eickhorst, K., Agouris, P., Partsinevelos, P.: Modeling and comparing change using spatiotemporal helixes. In: Proceedings of the Eleventh ACM International Symposium on Advances in Geographic Information Systems, New Orleans, Louisiana, USA, pp. 86–93 (2003)
[6] Meka, A., Singh, A.: DIST: a distributed spatio-temporal index structure for sensor networks. In: Proceedings of the 14th ACM International Conference on Information and knowledge management, CIKM 2005, Bremen, Germany, pp. 139–146 (2005)
[7] Pfoser, D., Jensen, C.: Trajectory indexing using movement constraints. GeoInformatica 9, 93–115 (2005)
[8] Chen, J., Meng, X., Guo, Y., Grumbach, S., Sun, H.: Modeling and predicting future trajectories of moving objects in a constrained network. In: Proceedings of the 7th International Conference on Mobile Data Management. MDM 2006, p. 156. IEEE Computer Society, Los Alamitos (2006)
[9] Dillenburg, J., Nelson, P., Wolfson, O., Yu, O., Sistla, A., McNeil, S., Ouksel, A., Xu, B., Ben-Arie, J.: Applications of a transportation information architecture. In: IEEE International Conference on Networking, Sensing and Control, vol. 1, pp. 480–485 (March 2004)
[10] Cao, H., Wolfson, O.: Nonmaterialized Motion Information in Transport Networks. In: Eiter, T., Libkin, L. (eds.) ICDT 2005. LNCS, vol. 3363, pp. 173–188. Springer, Heidelberg (2004)

[11] Wolfson, O., Jiang, L., Sistla, A., Chamberlain, S., Rishe, N., Deng, M.: Databases for Tracking Mobile Units in Real Time. In: Beeri, C., Bruneman, P. (eds.) ICDT 1999. LNCS, vol. 1540, pp. 169–186. Springer, Heidelberg (1998)

[12] Forlizzi, L., Güting, R., Nardelli, E., Schneider, M.: A data model and data structures for moving objects databases. In: Proceedings of the 2000 ACM SIGMOD international Conference on Management of Data, Dallas, TX USA, pp. 319–330 (2000)

[13] Güting, R., Schneider, M.: Moving Objects Databases. Morgan Kaufmann Publishers, San Francisco (2005)

[14] Rodríguez-Tastets, M.: Moving Objects Databases. Encyclopedia of Database Technologies and Applications, 377–385 (2005)

[15] Güting, R., Böhlen, M., Erwig, M., Jensen, C., Lorentzos, N., Schneider, M., Vazirgiannis, M.: A foundation for representing and querying moving objects. ACM Transactions on Database Systems 25, 1–42 (2000)

[16] Xie, R., Shibasaki, R.: A unified spatiotemporal schema for representing and querying moving features. SIGMOD Record 34, 45–50 (2005)

[17] Pfoser, D., Jensen, C.: Indexing of network constrained moving objects. In: Proceedings of the 11th ACM international Symposium on Advances in Geographic information Systems 2003, pp. 25–32 (2003)

[18] Hornsby, K., Egenhofer, M.: Modeling moving objects over multiple granularities. Annals of Mathematics and Artificial Intelligence 36, 177–194 (2002)

[19] Vazirgiannis, M., Wolfson, O.: A Spatiotemporal Model and Language for Moving Objects on Road Networks. In: Jensen, C.S., Schneider, M., Seeger, B., Tsotras, V.J. (eds.) SSTD 2001. LNCS, vol. 2121, pp. 20–35. Springer, Heidelberg (2001)

[20] Wolfson, O.: Moving objects information management: the database challenge. In: Halevy, A.Y., Gal, A. (eds.) NGITS 2002. LNCS, vol. 2382. Springer, Heidelberg (2002)

[21] Mark, D., Skupin, A., Smith, B.: Features, Objects, and Other Things: Ontological Distinctions in the Geographic Domain. In: Montello, D.R. (ed.) COSIT 2001. LNCS, vol. 2205, pp. 489–502. Springer, Heidelberg (2001)

[22] Smith, B., Mark, D.: Geographical categories: an ontological investigation. International Journal of Geographical Information Science 15(7), 591–612 (2001)

[23] Fonseca, F., Egenhofer, M., Agouris, P., Câmara, G.: Using Ontologies for Integrated Geographic Information Systems. Transactions in GIS 6(3), 231–257 (2002)

[24] Konstantinou, N., Spanos, D.-E., Chalas, M., Solidakis, E., Mitrou, N.: VisaVis: An approach to an intermediate layer between ontologies and relational database contents. In: Proceedings of the International Workshop on Web Information Systems Modeling: WISM 2006 (2006)

[25] Yuan, A., Mylopoulos, J., Borgida, A.: Building Semantic Mappings from Databases to Ontologies. In: Proceedings of AAAI (2006)

[26] Klein, M.: Combining and relating ontologies: an analysis of problems and solutions. In: Workshop on Ontologies and Information Sharing, IJCAI 2001, Seattle, USA (August 2001)

[27] Miller, H., Shaw, S.: Geographic Information Systems for Transportation, p. 480. Oxford University Press, Oxford (2001)

[28] Ding, Z., Guting, H.: Managing moving objects on dynamic transportation networks. In: Proceedings of the 16th international Conference on Scientific and Statistical Database Management (SSDBM 2004). SSDBM, pp. 287–296. IEEE Computer Society, Los Alamitos (2004)

[29] Saltenis, S., Jenson, C., Leutenegger, S., Lopez, M.: Indexing the positions of continuously moving objects. ACM SIGMOD 2000, 285–289 (2000)

[30] Gkoutos, G., Green, E., Greenaway, S., Blake, A., Mallon, A., Hancock, J.: CRAVE: A database, middleware and visualization system for phenotype ontologies. Bioinformatics 21(7), 1257–1262 (2004)

[31] Euzenat, J., Loup, D., Touzani, M., Valtchev, P.: Ontology alignment with OLA. In: Third EON Workshop. Third International Semantic Web Conference, pp. 333–337 (2004)

[32] Straccia, U., Troncy, R.: oMAP: Combining classifiers for aligning automatically OWL ontologies. In: Ngu, A.H.H., Kitsuregawa, M., Neuhold, E.J., Chung, J.-Y., Sheng, Q.Z. (eds.) WISE 2005. LNCS, vol. 3806, pp. 133–147. Springer, Heidelberg (2005)

[33] Fridman, N., Musen, M.: PROMPT: Algorithm and tool for automated ontology merging and alignment. In: Proceedings of the Seventennth National Conference on Artificial Intelligence and Twelfth Conference on Innovative Application of Artificial Intelligence, pp. 450–455. AAAI Press / The MIT Press (2000)

[34] An, Y., Mylopoulos, J., Borgida, A.: Building semantic mappings from databases to ontologies. In: Proceedings, The Twenty-First National Conference on Artificial Intelligence an the Eighteenth Innovative Applications of Artificial Intelligence Conference. American Association for Artificial Intelligence (AAAI) (2006)

[35] Niles, I., Pease, A.: Towards a Standard Upper Ontology. In: Welty, C., Smith, B. (eds.) Proceedings of the 2nd International Conference on Formal Ontology in Information Systems (FOIS-2001), Ogunquit, Maine, October 17-19, 2001 (2001)

Applications

An Operational Real-Time Ocean Sensor Network in the Gulf of Maine

Neal R. Pettigrew[1], Collin S. Roesler[2], Francois Neville[3], and Heather E. Deese[1]

[1] School of Marine Sciences, University of Maine, Orono, ME 04469 USA
[2] Bigelow Laboratory for Ocean Sciences, West Boothbay Harbor, ME 04575, USA
[3] Spatial Information Engineering, University of Maine, Orono, ME 04469 USA

Abstract. The Gulf of Maine Ocean Observing System (GoMOOS) was established in the summer of 2001 as a prototype real-time observing system that now includes eleven solar-powered buoys with physical and optical sensors, four shore-based long-range HF radar surface current systems, circulation and wave models, satellite observations, and hourly web delivery of data.

The Gulf of Maine is a harsh operational environment. Its winter storms pose severe challenges including the build up of sea ice on buoy sensors, superstructure, and solar panels, and in summer its productive waters present severe biofouling problems that can affect the optical sensors. The periods of most difficult operations often coincide with periods of greatest data value in terms of marine safety, search and rescue, and monitoring biological productivity. GoMOOS scientists and engineers continue to refine system designs and operational procedures to moderate the environmental stresses on the sensors and systems.

Keywords: Ocean Observing Systems, Sensor networks, real-time, GoMOOS, neural networks, ocean optics, CODAR, Gulf of Maine.

1 Introduction

The Gulf of Maine Ocean Observing System (GoMOOS) is a comprehensive prototype integrated coastal ocean observing system. It serves a broad array of real-time oceanographic and marine meteorological data and data products to scientists, state and federal regulators, the National Weather Service, both the US and Canadian Coast Guards, the National Data Buoy Center, educators, regional natural-resource managers, the Gulf of Maine fishing and maritime industries, local airports and airlines, sailors, and the general public. As the first of the truly operational coastal observing systems, GoMOOS has also served as a proving ground for procedures, protocols, and technologies, as well as an example of meaningful integration of sensors, platforms, and predictive models.

In addition to the hourly operational data delivery, GoMOOS provides an archive of data and model output that are significantly advancing the scientific understanding of the Gulf of Maine as a physical and ecological system. Over the nearly seven years of operation, the GoMOOS data have not only revealed the seasonal and interannual

S. Nittel, A. Labrinidis, and A. Stefanidis (Eds.): GSN 2006, LNCS 4540, pp. 213–238, 2008.

variability of the circulation and physical properties of the Gulf of Maine, in some locations they have provided the first meaningful baseline data. In addition, the Go-MOOS optical measurements have provided the first long-term concurrent measurements of chlorophyll fluorescence, spectral phytoplankton absorption, and particle size that has permitted novel time series records of phytoplankton blooms and species succession. The system also provides a "window on the Gulf" to educators, students, and the general public that performs an important educational function.

GoMOOS is an integrated ocean observing system that can be thought of as consisting of four major subsystems: the data acquisition subsystem; the data handling, processing, and archiving subsystem; the system of numerical nowcast and forecast models; and a web-based data distribution/presentation subsystem. The acquisition system includes a real-time buoy array, an array of land-based long-range Coastal Ocean Dynamics Applications Radar (CODAR) installations, and a satellite receiving station. The data handling system includes real-time QA/QC algorithms, data processing, calibration, and data archives that include sensor inventories, deployment histories and calibration records in addition to the data records themselves. The numerical modeling system [1] consists of an application of the Princeton Ocean Model (POM) for circulation and hydrographic conditions, and a high-resolution SWAN wave model. Both models use output from the Eta mesoscale atmospheric forecast model run by the National Center for Environmental Prediction (NCEP).

1.1 The Oceanographic Domain of the Gulf of Maine

The GoM is a complex and very productive marine ecosystem. The level of primary (phytoplankton) production, which forms the base of the marine food chain, is high relative to other continental shelf and marginal sea environments. Fisheries production in the GoM is also high. In particular, Georges Bank, which separates the interior GoM from the northwestern Atlantic Ocean, is one of the most productive fishing regions in the world. These high levels of production are believed to be due to a combination of nutrient delivery via the deep inflow of nutrient-rich slope waters through the Northeast Channel, which cuts between Georges Bank and Browns Bank, and strong tidal mixing that effectively mixes the nutrients up into the lighted (euphotic) zone were they are available to fuel phytoplankton blooms.

Under modern climate conditions, the Gulf of Maine is a region of strong physical, chemical, and biological gradients. During summer conditions there is a strong contrast in upper water properties from warmer, fresher, and lower nutrients in the southwest to colder, saltier, and higher nutrients in the northeast (e.g. Pettigrew et al., 2005, www.gomoos.org). There is also a clear southwest to northeast increase in tidal amplitude, and thus the degree of vertical mixing. A surface temperature front, which trends offshore from mid coastal Maine in the vicinity of Penobscot Bay, often develops that in summer separates the warmer surface waters of the southwestern region from the colder waters of the northeastern gulf [2].

The strong southwest-to-northeast gradients in physical processes and water properties in the GoM are mirrored in the seasonal patterns of phytoplankton blooms and the species composition. Spring blooms propagate from southwest to northeast, following the pattern of thermally-induced stratification, while fall blooms propagate from northeast to southwest in response to cooling and convection. Under present climatic conditions, we find within the Gulf of Maine, the northern coastal geographical limit of many temperate species and the southern coastal limit of many boreal species [3]. The affected temperate species are limited to the southwestern gulf, and conversely, the affected boreal species are limited to the northeastern GoM. These circumstances make the GoM an area that is biologically sensitive to the climate change signals since modest changes temperature can result in large swings of the species composition (ranging from phytoplankton species to commercial fish species) within the gulf. Thus the GoM may be expected to be a region in which the early effects of climate change will be manifest, and a region in which ocean observing efforts may foretell events of great ecological and commercial significance.

The circulation and physical processes in the GoM are dominant factors that determine the overall character of the ecosystem. In effect, the physical oceanography of the GoM is the fluid-mechanical regime within which the biochemical components of ecosystem are embedded. However, despite recognition of the importance of the physical regime, little is known in detail about even the most fundamental physical processes in the GoM, including its circulation.

The GoM is known to have a cyclonic general circulation pattern [4,5], and its shelf regions are characterized by a complex, variable, and interconnected coastal-current system that is best developed in the summer season [5,6]. A schematic diagram of the near-surface summer circulation, shown in Fig. 1 [6], depicts a pair of cyclonic (anti-clockwise) gyres over the basins in the eastern Gulf, and a partial separation from of the coastal current from the shelf at a mid-coast location in the vicinity of Penobscot Bay. Even this somewhat complex pattern is highly simplified in both time and space.

Recent moored current measurements and hydrographic surveys from the Ecology and Oceanography of Harmful Algal Blooms (EcOHAB) experiment in the Gulf of Maine show marked interannual variability in the degree of separation versus through-flow that occurs along the coast [6]. This seasonal and interannual variability in the connectivity of the eastern (EMCC) and western (WMCC) branches of the coastal current system is expected to have far-reaching consequences with regard to the transport of nutrients, planktonic larvae, harmful algal blooms, and coastal pollution. All of these fluxes are critical factors in determining both short and long-term variations in the state of the GoM ecosystem. GoMOOS moorings I and E, which are, respectively, in the EMCC and WMCC, have revealed strong seasonal component to this connectivity. The two branches generally merge each fall and separate each spring. The historic lack of long-term direct current measurements made within the Gulf of Maine Coastal Current (GMCC), and at key inflow and outflow locations near the open boundary, has significantly hindered our understanding of both the physical and biological oceanography of the GoM.

The surface inflow into the GoM of relatively fresh Scotian Shelf water (SSW) from the Atlantic seaboard of Nova Scotia, and the deep inflow through the Northeast Channel of relatively warm, salty, nutrient rich slope waters (SLW) are the two most important inflows into the GoM. The large buoyancy input of the SSW accounts for more of the annual freshwater budget of the GoM than the combined inflow of all the rivers that drain into its confines. The density contrast between these relatively fresh surface and intermediate SSW waters with the deep salty slope SLW waters survives

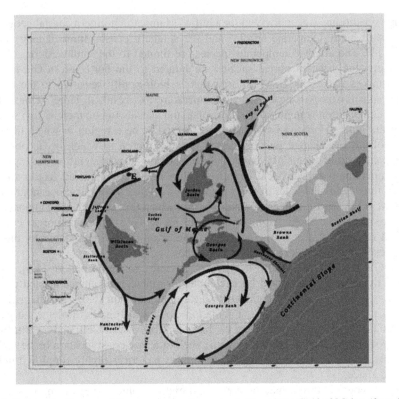

Fig. 1. Schematic diagram of the summer surface circulation of the Gulf of Maine (from Pettigrew et al., 2005)

the vigorous tidal mixing and winter convection in of the eastern GoM. The geostrophic adjustment processes in response to these persistent density contrasts engender the cyclonic general circulation pattern of the GoM. The monitoring of these inflows has long been recognized as a prerequisite to any credible Ocean Observing System, as well as to the understanding the large interannual and decadal variability that characterize the hydrographic structure and the fisheries yields of the GoM.

The lack of biological time-series measurements has been even more limiting than the paucity of physical time-series measurements. Prior to the implementation of the GoMOOS, there had been no significant time-series measurements that reflect phytoplankton biomass, other than those associated with shipboard programs in which short time series at a single station or quasi-synoptic geographical surveys were implemented (e.g. GLOBEC, ECOHAB), ;and remotely-sensed chlorophyll, which in this region of high riverborne colored dissolved organic matter (CDOM) has high error, particularly near the coast. The GoMOOS array has, for the first time, enabled hourly estimates of phytoplankton biomass, and photosynthetically available radiation (PAR) at multiple buoy locations within the euphotic zone.

Rivers serve important roles in the Gulf of Maine, from the perspectives of the quantity of freshwater entering the system, forcing of buoyancy-driven coastal flow, and the inorganic nutrient and organic material fluxes into the coastal waters. These fluxes are important factors that influence coastal algal blooms (including harmful algal blooms of the genus Alexandrium, [5,6] and the sequestering of particulate and dissolved carbon and nitrogen in the deep basins of the Gulf of Maine [7]. The optical packages that we have deployed on selected moorings provide us with the capability to determine both the concentration of the particulate and dissolved matter, and the composition and size distribution of that material.

2 The Ocean Observing System

2.1 GoMOOS Real-Time Buoy Designs

The GoMOOS Data Buoy System design arose from the central concept of a moderate sized, stable, unsinkable, solar-powered platform with real-time telemetry and onboard data processing capabilities. The GoMOOS design incorporates many significant departures from the buoy designs used previously in the Gulf of Maine for year-round operation. Funding constraints and lean operational budgets dictated that the buoys be much cheaper and more compact than the three-meter discuss buoys deployed by the National Data Buoy Center (NDBC) in the GoM region. The smaller, lighter buoys allow much smaller (and more economical) vessels to be used for deployment and recovery. At the same time, while required to be smaller, cheaper, and lighter than the NDBC buoys, the GoMOOS buoys had to be capable of handling an order of magnitude greater data volume, and an order of magnitude more sensors, while still withstanding the rigors of the Gulf of Maine winters.

The "workhorse" of the GoMOOS buoy array is a multi-chinned two-meter discus buoy with flotation made of closed-cell Surilin foam. There is a central water-tight instrument well, made of aluminum, which houses the buoy electronics including the voltage regulation system, solar storage batteries, and the data-logger/controller. The buoy is designed to survive knockdown and compression due to forced submergence, it monitors its own position, and sends alarms if it detects it is off position, has a leak in the electronics well, or has data logging problems. The buoy is solar-powered, has dual cellular/iridium and GOES satellite telemetry systems for hourly data telemetry, and has room for expansion in its both power and electronic systems. Some of the design details of the GoMOOS buoy system have been previously described [7].

The basic GoMOOS buoy platform is used in two distinct configurations: the shelf mooring, and the basin mooring. The schematic diagrams of the GoMOOS shelf and basin moorings are shown in Fig. 2A and 2B. All buoys are mechanically identical and carry a surface sensor payload of anemometers (for wind speed and direction), air temperature, atmospheric pressure, and visibility (fog). A subset of the buoys (the optics-intensive buoys) also measure solar insolation.

The majority of the GoMOOS buoys are shelf moorings of the slack chain type shown in Fig. 2A. These mooring designs have proven very reliable when deployed in the coastal GoM. In water depths of 100 or less, the scope (or slack) of the mooring is provided by approximately 60 m of 5/8 inch mooring chain that connects the 5/16 inch jacketed steel mooring cable to a 2700 lb anchor. Under slack water low tide conditions, approximately 10 m of the chain is suspended above the bottom and the remaining 50 m lies on the bottom. During high water, high wave, or high flow conditions the chain pile is picked up by the buoy as needed to adjust to the lift and drag. An advantage of this simple mooring design is that the anchors are recovered with the buoy, and no acoustic release mechanism is required.

At the deeper offshore deployment sites we use the compliant mooring system shown in Fig. 2B. In this design, the scope is provided by elastic tethers that hold the buoy under tension and stretch in reaction to high waves, tides, or drag from strong currents. The chief advantage this design in the GoM is that the "watch circle" of the buoy around its anchor is reduced. In a region that is heavily fished, the reduced watch circle makes it less likely that draggers will work between the anchor and the buoy, and thus less likely that the buoy will be cut, dragged off station, and its sensors damaged. Since the fishing activity is greater in the offshore waters of the GoM, GoMOOS uses the elastic tether mooring systems in these locations. The elastic tether also makes the surface buoy behave in a more "spar like" fashion in a wave field: that is, with less rolling of the surface float. A disadvantage of the elastic tether design is that the tethers are expensive, need to be replaced approximately annually, and must be used in conjunction with expensive acoustic release mechanisms that leave the anchors on the bottom.

We have recently designed an inshore "mini" buoy that uses on the slack chain mooring configuration. This buoy has been designed to be deployed and recovered using small work boats of ~ 30 ft or greater, and to operate in the small bays and estuaries of the GoM where the storm waves are less than approximately 2 m. In terms of telemetry, data handling, and sensor payload the nearshore buoy is the equivalent of the full-size GoMOOS buoy; without the room for future expansion in the instrument (electronics) well. However it is designed to survive encounters with ice flows that are common in GoM estuaries, but virtually unheard of in the open gulf. The buoy schematic is shown in Figure 2C.

2.2 Local Sensor Networks

The GoM is a diverse physical and ecological environment and the monitoring system reflects this diversity. Never-the-less, each mooring is equipped with a standard suite of instruments in addition to the site -specific sensors dictated by the variable conditions around the GoM. Standard for all buoys are a meteorological package consisting of an R.M. Young wind sensor and a sonic anemometer (either Gill or Vaisalla) that measure wind speed and direction, an Aanderaa visibility sensor, a Sutra atmospheric pressure sensor, and a Campbell Scientific air temperature sensor. Wave parameters are estimated by an onboard Summit Tri-axis accelerometer. In-water sensors include a Seabird microcat measuring temperature and conductivity mounted on the base of the buoy at a depth of 1 meter, and an Aanderaa RCM-9 MKII current meter measuring current velocity at 2 meters. All moorings also additional carry Seabird microcat and/or seacats that measure temperature and conductivity at various depths. Typical configurations for shallow and deep moorings can be seen in the schematic diagrams of Figs. 2A and 2B.

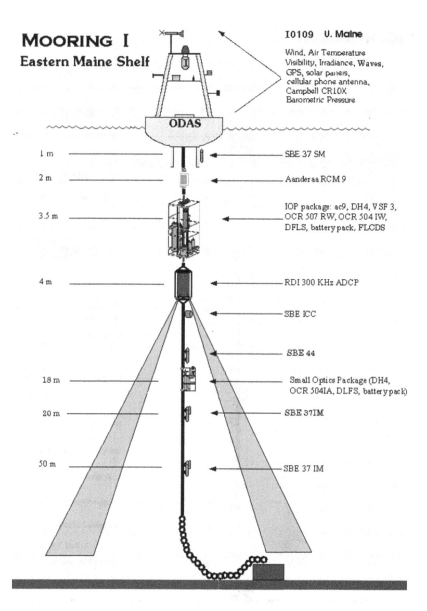

MOORING I

Eastern Maine Shelf

I0109 U. Maine

Wind, Air Temperature
Visibility, Irradiance, Waves,
GPS, solar panels,
cellular phone antenna,
Campbell CR10X
Barometric Pressure

ODAS

1 m — SBE 37 SM

2 m — Aanderaa RCM 9

3.5 m — IOP package: ac9, DH4, VSF 3, OCR 507 RW, OCR 504 IW, DFLS, battery pack, FLCDS

4 m — RDI 300 KHz ADCP

— SBE ICC

— SBE 44

18 m — Small Optics Package (DH4, OCR 504IA, DLFS, battery pack)

20 m — SBE 37IM

50 m — SBE 37 IM

Fig. 2A. Schematic diagram a solar-powered GoMOOS Shelf Buoy for use in approximately 100 m depth in the open Gulf of Maine. Scope of the mooring is provided by a chain pile on the bottom. Subsurface sensors send data up the mooring wire using inductive modem technology.

MOORING M

Jordan Basin

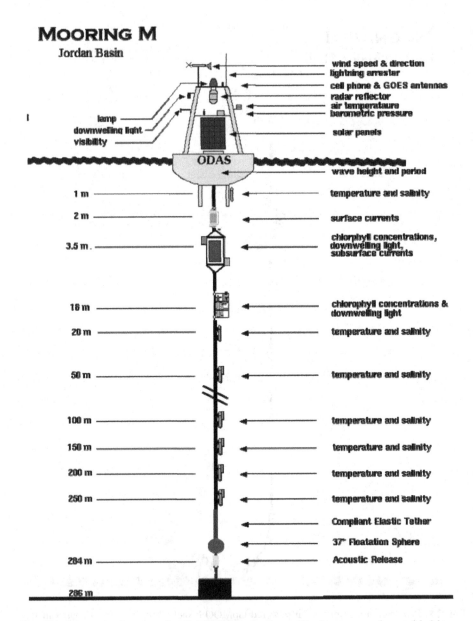

wind speed & direction
lightning arrester
cell phone & GOES antennas
radar reflector
air temperature
barometric pressure

lamp
downwelling light
visibility

solar panels

ODAS

wave height and period

1 m — temperature and salinity

2 m — surface currents

3.5 m — chlorophyll concentrations, downwelling light, subsurface currents

18 m — chlorophyll concentrations & downwelling light

20 m — temperature and salinity

50 m — temperature and salinity

100 m — temperature and salinity

150 m — temperature and salinity

200 m — temperature and salinity

250 m — temperature and salinity

Compliant Elastic Tether

37" Floatation Sphere

Acoustic Release

284 m

286 m

Fig. 2B. Schematic diagram of a GoMOOS Basin Buoy. The buoy scope is provided by an elastic tether near the bottom that reduces the watch circle of the mooring. Subsurface sensors send data up the mooring wire using inductive modem technology

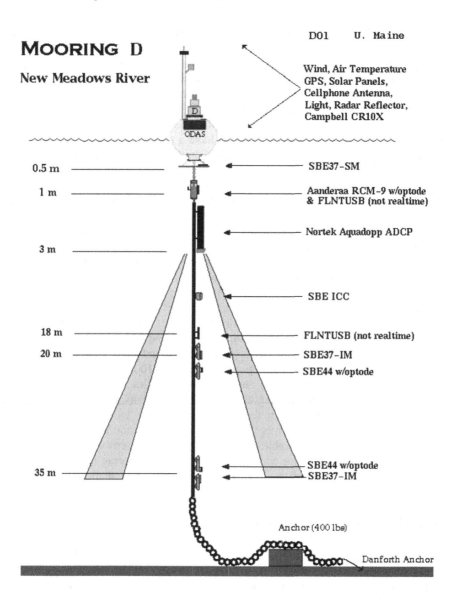

Fig. 2C. Schematic diagram of a GoMOOS Estuarine Buoy. This buoy is the shallow water version of the Shelf buoy of Fig. 2A. The buoy mast is only 2 m above the water line, and the 500 lb. buoy is capable of being deployed from a small work boat. Subsurface sensors send data up the mooring wire using inductive modem technology.

In all GoMOOS buoy/mooring designs, meteorological sensors (wind-speed and direction, air temperature, atmospheric pressure, visibility, and incident light) and the shallow subsurface sensors within 4 m of the sea surface are hardwired to a Campbell Scientific data logger that is housed within the buoy's instrument well. Seabird Instruments temperature and conductivity sensors are attached at 0.5m or 1 m below the surface on the cross members that support the buoy legs. Surface currents are measured at 2m depth by an Aanderaa RCM9 MKII Doppler current meter (1m on the nearshore buoy). Optically-equipped buoys have a cluster of optical sensors (which include chlorophyll and CDOM fluorometers, multi-wavelength absorption and attenuation, meters (ac9), backscattering sensors, and spectral upwelling radiance and downwelling irradiance radiometers) at 3.5 m. Most buoys have downward-looking acoustic Doppler Current profilers to measure subsurface currents. The majority of the Doppler profilers are 300 kHz RD Instruments "Workhorse" Doppler Profilers with a profiling range of approximately 100m at 4 m vertical resolution. Buoys at deep locations are equipped with 150 kHz or 75 kHz RD Instruments profilers that have ranges of approximately 200 and 400 m, respectively. The shallow nearshore buoys are equipped with Nortek 400 kHz profilers, or RD Instruments 600 kHz profilers that have 80-60 m range and are set with 1 m or 2m vertical resolution.

Data from subsurface sensors deeper than 4 m, including temperature and conductivity, dissolved oxygen (at select mooring locations) and deep optical packages (chlorophyll fluorometers and downwelling spectral irradiance sensors), are transmitted up the jacketed steel mooring cable via the Seabird inductive modem system. Use of this system makes possible the deployment of up to 100 addressable sensor packages without requiring the running underwater data cables. Instead, the sensors are inductively coupled to the mooring cable itself. The sensors are sequentially polled each hour, and the data provided to the data logger and buoy telemetry system.

The inductive modem system is very flexible and an important component of the array architecture. Electrical cables are a major source of sensor failure in moored applications as well as a major contributor to hydrodynamic drag in region of strong currents. Through the use of the inductive modem system, a shallow sensor failure can be remedied by a diver who unclamps the sensor package and replaces with another, thus avoiding a complex and expensive mooring recovery. In addition, changes in the depths of sensor deployment only involve attaching the sensor somewhere else on the wire rather than remanufacture of the electrical cable and connectors. Each of the sensor packages are internally powered and also provide internal recording. Thus in case of failure or intermittency of the inductive modem system the data archive will be intact after mooring recovery, although the data will not be available in real time.

2.3 The GoMOOS Buoy Array

The GoMOOS buoy array is shown in Fig. 3. The red dots with blue letters show the locations of the twelve GoMOOS buoy locations within the GoM. All but buoy K are active buoys: Buoy K was relocated (to station N) due to shifting scientific priorities. Reference to Figure 1 shows that many of the buoys are located within the Gulf of Maine Coastal Current System (GMCC) [6] Buoys L and N monitor, respectively, the inflows into the Gulf of Maine from the Scotian Shelf and the North East Channel that

Fig. 3. The Gulf of Maine Ocean Observing System. Buoys are red dots, CODAR stations are lavender "bull's eyes", green diamonds show the NOAA island meteorological stations, and amber diamonds show NOAA meteorological buoys.

were discussed earlier. Buoy M, located in the Jordan Basin, monitors the seasonal inventory of the nutrient-rich SLW in the interior of the GoM, which contributes to the high productivity of this famous fisheries region.

Buoys L, I, E, and B, all located in the Gulf of Maine Coastal Current (GMCC) system, collectively represent the first buoys that have gathered long time-series data at multiple sites within the GMCC. They have now all been in nearly continuous operation since July of 2001. Buoys A, C, F and J monitor the mouths of four GoM major estuarine embayments that are prominent in the GoM from the standpoints of commerce, fisheries, aquaculture, and recreation.

Associated with the eleven active GoMOOS buoy sites, are twenty two buoys and twenty two complete sets of instrumentation. At approximately six-month intervals (Spring and Fall) the entire array of buoys and instruments are exchanged for a set that have been refurbished, tested, repaired and calibrated in the interim since their previous deployment. This rotation of sensors and buoys has kept the GoMOOS data streams among the most reliable of the operational ocean sensor arrays.

The data quality assurance program within the Physical Oceanography Group at the University of Maine, which runs the "in water" GoMOOS sensor array is a program of continual quasi-realtime evaluation and validation. Detailed histories are kept of sensor performance, repair, and recalibration. Signatures of failure modes for the various sensors have been identified over the years of operation since 2001, and real-time performance is consistently evaluated and compared with these models of failure in order to flag suspect data in near realtime. With each passing deployment year, the QA/QC procedures become more skillful and further automated.

2.4 The GoMOOS Optics Program

The GoMOOS optics program uses two moored optics packages: the "phytoplankton biomass and production" packages and the larger "Ocean Color packages." The phytoplankton packages consist of WET Labs digital chlorophyll fluorometers (DFLS) and Satlantic 4-channel downwelling irradiance sensors (OCR504I), used to estimate phytoplankton biomass and photosynthetically available radiation (PAR), respectively. The combination of phytoplankton biomass and light measurements are sufficient for calculating primary production to first order [8]. These instrument packages are powered and controlled by a WET Labs DH4 data handler that logs ~1 min of 6 Hz data, and computes statistics of the burst sampling which are relayed to the Campbell data logger. Upon recovery, the full data set is archived and reprocessed applying post-recovery calibrations and corrections [9]. The ocean color packages have all the instruments of the smaller package, but also incorporate a WET Labs AC9 nine-wavelength absorption and attenuation meter, a WET Labs ECOvsf, backscattering sensor, a WET Labs FLCDS CDOM fluorometer, and a Satlantic OCR507R 7-channel upwelling radiance sensor. Each of the optical moorings carry phytoplankton packages at 3 and 18m and four-channel Satlantic OCR incident spectral irradiance sensors on top of the buoy. Moorings E and I have the full ocean color packages at 3 m and three additional channels on the incident irradiance sensors to complement the radiance sensor.

The most obvious problem with deploying optical sensors for extended periods in the ocean is biofouling. We have employed a number of strategies for minimizing biofouling, but it is also important for us to be able to identify, in real time, when biofouling is a serious problem and to create strategies for correcting the data, either in real time or after sensor recovery. The sensors are configured with combinations of copper shutters that lie over optical faces until the instrument turns on, copper tape on surfaces surrounding the optical heads to prevent macrofaunal growth that might impede the sensor head (or worse, prevent the shutters from opening), and copper tubing on the flow through instruments. The copper dissolves in seawater, creating a toxic layer of water over the sensors while they are in sleep mode; once the sensors turn on, the shutters rotate away from the sensing head, or in the case of the flow through instrumentation, the water is flushed away by a pump prior to data collection. We have found generally these strategies to be effective, even during a 6-month deployment during the most productive portion of the year; particularly for the algal growth on chlorophyll fluorometers (Fig. 4). We do find some bacterial growth on the optical windows, particularly on the absorption-attenuation meter. For this reason, it is particularly important that we are able to identify when this is occurs in real time. The protocols for real-time and post-recovery calibration have so have been described elsewhere [10].

Each of the optical moorings (B, E, F, I and M) are equipped with multichannel downwelling irradiance sensors to measure the spectral solar radiation incident on the ocean surface and penetrating to 3 and 18m. From these observations we can compute the photosynthetically available radiation (by integrating over the visible spectrum) and compute the depth of the euphotic (or lighted) zone in which phytoplankton are actively photosynthesizing. The combination of phytoplankton biomass, euphotic

Fig. 4. Underwater photograph of WET Labs fluorometer, after 5 months in the water., Anti-biofouling copper tape has kept algal growth from affecting the optical window even thought it was deployed without a shutter. *Photo by Steve Karpiak, MER.*

depth and depth-resolved PAR provides the inputs to calculating integrated primary production [11]. In addition to observations of phytoplankton biomass, which are available on moorings A, B, E, F, I, and M, moorings E and I are also equipped with more complex optical packages that measure all the inherent and apparent optical properties that are found in the radiative transfer equation that describes the ocean color measured by the satellite-based SeaWiFS and MODIS sensors. The inherent optical properties (spectral absorption, attenuation and backscattering) describe the innate optical properties of the particulate and dissolved matter in the ocean, while the apparent optical properties (the ratio of the spectral upwelling radiance to the down-welling irradiance) describe the light field in the ocean, and ultimately the light that leaves the ocean and is visible by eye or by satellite sensors. These apparent optical properties are a function of the incident solar radiation and the inherent optical properties. Thus, using inverse modeling [11] ocean color observations can be used to predict the concentration and composition of the material in the ocean.

Finally, the sensor packages on E and I also allow us to provide calibration and validation of the satellite-based ocean color sensors. Measurements of the spectral upwelling radiance, which is akin to that measured aboard the satellites, provides the "sea truth" for the satellite observations once the atmospheric effects are removed. The moored optics program also provides validation, in that the system measures a range of ocean color products that NASA provides including chlorophyll concentration, dissolved, particulate and phytoplankton absorption, and backscattering. This capability allows the evaluation of the NASA products in a coastal environment, which has proven to be a major challenge to ocean color science.

2.5 The HF Radar Array

Use of radio waves in the HF and VHF bands for surface current measurement has been under investigation for several decades. The underlying principle is based on a phenomenon known as Bragg scattering. When the radio waves are transmitted over the wavy air/sea interface, the transmitted radio energy will be scattered directly back to its source when the radio signal scatters off a wave that is exactly half the transmitted signal wavelength (λ), and that wave is traveling either directly toward or away from the transmitter. Since scattering from the other waves occurs in all directions, the received (back scattered) signal comes overwhelmingly the Bragg wave of specific wavelength wave traveling in a known direction.

The HF radar system used in GoMOOS is the long-range Coastal Ocean Dynamics Applications Radar (CODAR) system, manufactured by CODAR OS. For a long-range CODAR (nominal 5 MHz transmission) a λ of ~30 m is the Bragg wavelength. Since the speed of wave is a function of its wavelength alone for water depths greater than $\lambda/2$ (deep water waves), the returned radio signal contains a Doppler frequency-shift due to it scattering from the moving sea surface. Once the wave speed is subtracted from this Doppler shift, the remainder is due to the movement of the surface itself; that is, due to the surface currents. If two or more transmitters and/or receivers cover the same region, radial speeds can be geometrically transformed into orthogonal horizontal velocity components, and the vector velocity of the surface currents can be determined.

The GoMOOS long-range CODAR array is a potentially exciting element of the observing system capable of widespread, remote surface-current measurements from a limited number of shore-based radio wave transceivers. These data are of increasing importance to several GoMOOS user groups including those dealing with search and rescue, contaminant transport, commercial shipping, recreational boating, and larval transport. The first three of the GoMOOS shore-based CODAR units have been installed at Wood Island in southern Maine, Greens Island on mid-coast Maine, and at Cape St. Mary in NS, Canada (see Fig. 3). A fourth site is being installed at a Coast Guard station on the southern extremity of Grand Manan Island, NB, near the mouth of the Bay of Fundy, and it is expected to shortly be operational.

The long-range CODAR has a nominal daytime range of 180 km, so that three or four units are capable of providing coverage over nearly the entire GoM (~400 x 200 km) at a spatial grid size of approximately 6 km. Regions in which signals of adjacent CODAR systems overlap (at sufficient angle), correspond to regions in which full two-dimensional surface currents can be obtained. One drawback of the long-range 5 MHz system (between the AM and FM radio bands) is that it is susceptible to ionospheric radio interference and low signal to noise ratios. As a result of this interference, substantial day-night variation in the range of the system occurs, with the range often dropping by 50% or more during the night. When this happens, the radials from the installations do not have much overlap and the area of vector current coverage can be very limited. One solution to this problem is to shorten the distance between CODAR stations. However, the process of seeking approval from local authorities and landowners has proved to be an arduous one. In addition, the funding of the GoMOOS systems has never been adequate to fund expansion of the initial four-unit CODAR infrastructure.

Because of the highly variable CODAR radial overlap and vector coverage in the GoM, and the very strong tidal variability it has been challenging produce useful surface current fields. The inability to consistently receive data throughout the tidal cycle often reduces the data to a gappy series of realizations that can not be properly averaged to give a consistent picture of the general circulation patterns. Because of these limitations, we have begun to apply Artificial Neural Networks in order to fill in missing data values to provide tidally-averaged surface current maps, and to work toward the goal of short-term, wide area predictions of the GoM surface current fields.

2.6 Applications of Neural Network Models to Sensor Array Data

Artificial Neural Networks (ANNs) are parallel arrangements of simple processing units that are loosely analogous in structure and function to a biological central neuron. ANNs are comprised of a large number of these neurons in multiple layers. These neurons connect input and output through a collection of weights and transfer functions designed to minimize the differences between predictions and realizations (data). By adjusting the weights, the ANN can be "trained" to make very accurate predictions. ANNs that adjust their weights automatically can "learn" to make predictions.

Back propagation neural networks, in particular, are powerful forecast tools. They are called back propagation networks because of the learning algorithm they use, in which error corrections propagate backward from the output layer to the input one, i.e. in the direction opposite to that of signal spreading during the normal network operation, during the weight adjustment process. The weights converge to their "trained" configuration by minimizing the network output's mean squared error.

With suitable inputs, Artificial Neural Networks are and effective method for filling in missing values and predicting future values of geophysical time series. The great advantages of ANN models are that they are statistical and relational rather than dynamical in nature, and they do not generally depend upon the existence of a simple linear relationship between input (forcing) and output (response). Thus physical-mathematical equations governing the process are not required for the forecast to be made. All that is required is a set of input data that are correlated with the desired output, and sufficient examples of the "correct answer" (observations of response) to train the model.

In order to develop an ANN capable of predicting the complex currents in the Gulf of Maine, we began first by working on models to nowcast and forecast surface currents measured by in situ current meters suspended 2 m beneath the GoMOOS buoys. These data are of high quality with a high signal-to-noise ratio; an attribute that is crucial for effective training of the ANN model.

A neural network model was developed for the prediction of missing and future surface current values [12]. The inputs to the nowcast model were measured local wind vectors from 2 and 3 hours prior (there is a lag between wind forcing and current response), tidal current predictions based on harmonic analysis of previous current meter data, measured surface currents 1, 2, and 4 hrs earlier; 16m current measurements 1 and 0 hrs earlier; and a mean of the previous tidal cycle (~12.5) hours. The ANN is extremely successful at predicting the currents. In fact, it is more accurate than the predictions of GoM dynamical numerical circulation that use the full nonlinear equations of motion [1].

Figure 5 shows over plot comparisons of the ANN nowcast and the observed vector current components. The vector correlation coefficient between nowcast and observed currents was ~0.95 for most of the moorings, which was generally 25% higher than for the tidal prediction alone. Forecast skill of the model was essentially the same as the nowcast skill out to 3 hours into the future. Predictions beyond 3 hours had significantly lower correlation coefficients. For example, vector correlation between currents predicted 12 hours into the future and the observed currents were 0.7-0.8. An example is shown in Figure 6.

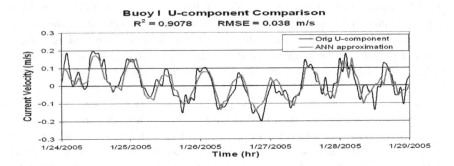

Fig. 5. Comparison plot of observed surface currents(black) and the nowcast predictions of a neural network model (grey). Model inputs were measured surface and deep currents 1, 2, and 4 hrs earlier; wind 2 and 3 hours earlier; a Tidal prediction and a mean of the previous12.5 hours.

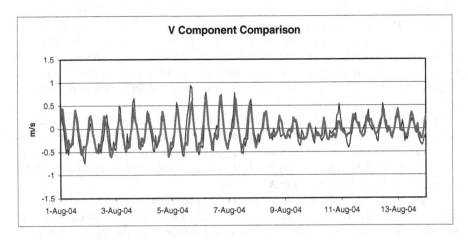

Fig. 6. Comparison plot showing, for a two-week period, observed surface currents and the ANN prediction 12 hours into the future. The thin dark line represents the observed currents and the thick blue line represents the predicted values. The correlation coefficient based on 4000 hours of comparison is 0.78.

Predictions of CODAR data fields are more problematic than predicting direct surface-current measurements from the GoMOOS buoy array. The CODAR data have a lower signal-to-noise ratio, and the available inputs for the model fluctuate in time and space. Because the data availability was spotty, we were unable to provide tidal predictions, average currents for the previous tidal cycle, or even previous values at each location as inputs. In fact, we were limited to wind forcing and a varying collection of current vector values from "nearest neighbors" in the CODAR field, and the radial values in regions where vectors where not available due to the absence of overlapping signals from adjacent CODAR units. In order to cope with the variable inputs, we developed a weighting system that produces a spatially-weighted value for the nearest neighbors so that unique ANNs need not be developed for each of the myriad possible input configurations.

The skill of the ANN-CODAR nowcasts has been very encouraging. Under favorable conditions, when some neighboring values are available, the vector correlation values between nowcast and observed values are approximately 0.9. Visual inspection of the plots of predicted and observed values shows that the discrepancies between observed and predicted values are often concentrated in episodes in which the CODAR data shows large values. Since the CODAR data are known to be noisy, the possibility exists that the ANN prediction is closer to the true value in some instances, and may prove useful as an automated quality control procedure. When large discrepancies exist between nowcast and observation, further assessment of the quality CODAR fields may be required.

An application of the ANN nowcast to the spatial CODAR surface current map is shown in Figure 7. Panel A shows the vectors from the CODAR vector data. Spatial gaps in the data are caused by reduced range so that radials do not overlap. The map in Panel B shows considerable improvement in coverage using winds, nearest neighbors, and radials as inputs to the ANN model. Although this example shows tremendous improvement, it represents relatively favorable conditions that are not always met, and further improvements in the model are required for a truly operational product.

The next logical step in the process of producing complete hourly surface current maps, is to add the output from a dynamical numerical circulation model as an input to the ANN model. The numerical model output has the great advantage of being regular in both space and time. When combined with wind forcing, CODAR radials, and weighted nearest neighbor (possibly including the directly measured moored current records), the ANN will improve upon the already reasonable surface current prediction of the dynamical model. Comparison of surface current meter records and numerical model output indicates that most dynamical models over respond to wind forcing. A hybrid dynamical-neural model is expected to improve the wind-driven predictions significantly.

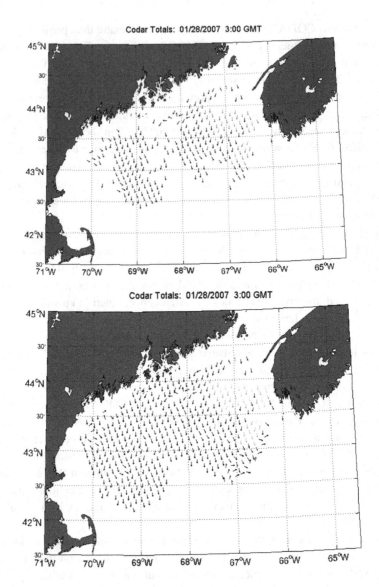

Fig. 7. A, B. Upper plot (A) shows surface current vectors observed by CODAR. Lower plot (B) shows the ANN nowcast of the missing vectors that satisfy the model requirements. The inputs to the model are: Radial Doppler data, weighted-average of neighboring values, and winds.

3 GoMOOS Operations

The data return for the GoMOOS ocean sensor array has averaged approximately 90% over the first five years of operation. This unusually high rate of data return is due in large part to our operational six-month duty cycle on all equipment. We have 22 buoys for 11 locations, and 22 complete sets of instrumentation. The identical sets of sensors are rotated in and out of service on a six-month deployment/maintenance schedule.

Despite the high success rate achieved, some persistent and recurrent technical problems have occurred. Some of these problems arise from the harshness of the GoM environment, and others from the familiar problems that have plagued oceanographic field programs for decades.

While the buoy itself is solar powered, the majority of the sensors are powered by internal batteries. As the instruments age, and are battered by storms and at-sea operations, they seem more prone to transient high electrical current drains. Our surface current meters have been particularly prone to these problems. In response to the battery packs of our current meters dying prematurely, we have added auxiliary battery packs inside the buoy instrument well. This strategy has not entirely eliminated the problem although we now power the surface current meters with power packs that have four times the expected power requirements. In addition to high current drains, performance of many sensors occasionally suffers from substandard batches of batteries from various manufacturers.

Another common mode of failure has been fatigue of electrical cables that connect surface and near-surface sensors to the buoy system. This problem will probably never entirely disappear, but it has been significantly improved with very careful cable strain relief, by using Kevlar-reinforced electrical cabling, and by routinely replacing all cables annually.

The extremely harsh environmental conditions in the GoM in winter has caused wave damage to solar panels that power the buoy system, bearing failures on anemometers, freezing of the anemometer rotors, and icing of the buoy superstructure including the solar panels (which causes low power generation). Solar panel breakage has been eliminated by reinforcing the backs of the panels with aluminum plates and foam cushioning, and all buoys have been fitted with sonic anemometers that are less prone to damage from freezing spray. The icing problems in winter remain a serious concern. The buoys were designed to perform with several hundred kg of sea ice build up, but under very cold storm conditions the ice build up exceeds design limits. We are presently considering a retrofit of the winter buoys in which the solar panels will be raised to reduce ice build up, and the addition of more ballast below the water line to increase buoy stability under heavy icing conditions.

Early in the operation of GoMOOS, we experienced significant damage and service interruptions due to fishing activity and barge traffic that have occasionally damaged sensors and moved buoys off station. The gear conflict problems with the fishing and marine transport industries have decreased with time as the fleets have become accustomed to the presence of the buoys.

As described earlier, biofouling of the optical sensors has at times been a serious problem in GoMOOS [10]. The problem has been reduced significantly through the use of copper tape, tubes, and shutters, although the mechanical reliability of the copper shutters is a continuing area of research and development. Biofouling is not a serious problem for most of the other GoMOOS sensors. The acoustic Doppler current meters are essentially immune to even extreme biofouling because the sound waves travel easily through the attached fouling organisms, and because the acoustic current meters are remote sensors that make measurements a distance of meters-to-hundreds-of-meters from the sensor surface. The temperature and conductivity sensors, while potentially vulnerable, are protected by poison tubes supplied by the manufacturer.

3.1 Preliminary Scientific Results from the Sensor Array

An example of the impact of the GoMOOS system on our ability to study the variability in the GoM is shown in Fig. 8. Prior to the implementation of the GoMOOS system, there were very few long-term data records in the GoM. In fact, there were no time series from the Gulf of Maine Coastal Current (GMCC) system that exceeded one year in length. The data shown in Figure 8 are hourly salinity values for five years from August 2001 through August 2006 at 50 m depth at several locations in the GMCC system. The usual seasonal salinity progression at these locations at 50 m depth is a salinity minimum in May or June, following the annual snow melt and spring rains, and a subsequent rise of 1.0-1.5 PSU, peaking in late fall or early winter. However, after the June 2004 minimum, the salinity rebounded by less than half the usual seasonal rise and then fell again in response to an unusually high runoff the following spring. This pattern was observed at every GoMOOS measurement site from the surface down to 250 m depth (in the central GoM). As is evident in Figure 8, low salinity values in the GoM did not rebound until that late summer of 2006, resulting in significant low-salinity anomaly pattern that lasted for two years.

The fundamental cause of this widespread salinity anomaly is clearly not the high runoff, although the runoff was a contributing factor some six months into the anomaly event. From our discussions of section 1.1, one might expect that at the beginning of the low salinity anomaly in the fall, there was less inflow of salty slope water and a compensating increase in the inflow of fresher Scotian Shelf water (in order to conserve volume in the GoM). As shown in Figure 9, direct measurements of the salinity and transports in the deep waters of the Northeast Channel confirm that not only did the characteristic inflow of salty slope waters decrease, it actually reversed for the fall and winter of 2004 and again (although more weakly) in 2005 (Figure 9). This unprecedented outflow event resulted in a net outflux of salinity in the deep waters and (presumably) greater inflow of fresher waters near the surface from the Scotian shelf. The resulting massive freshening of the GoM in 2004 and 2005 represents a major perturbation to the normal ecological conditions that is presently under further investigation.

3.2 Future Ocean Observing in the GoM: A Five-Year Horizon

Ocean sensor and ocean platform technologies are in a period of rapid development. The operational systems (such as GoMOOS) now in the water were not feasible one decade ago. Assuming reasonable funding levels, we can expect continued rapid advancement in the capabilities of operational ocean sensor networks within the next five years. These advances will include significant growth in the areas of real-time biochemical sensors, moored profilers, sensor miniaturization, lower power consumption, autonomous mobile sensor platforms, high speed data telemetry technologies, increases in onboard data processing capabilities, advances in data visualization techniques, and artificial intelligence.

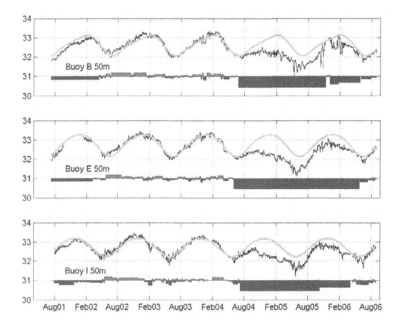

Fig. 8. Observed records of salinity observations (black), least-squares fit of annual and semi-annual periodicites to "normal" years (grey), and average strength of anomalies (lt. grey >0,dark grey >0) at 50m depth at three coastal shelf buoys (see Fig. 3 for locations) showing the gulf-wide low salinity anomaly of 2004-2005.

In situ moored nutrient sensors are presently capable of moored operation. These sensors fall into two categories: those using wet chemistry and those that don't. In general the reagent-based sensors have four issues that limit the practical deployment duration: reagent capacity, reagent stability, power, and biofouling. It is presently a challenge to achieve more than a few months deployment. Biofouling often the most severe limitation is shallow regions, but in deeper waters below the euphotic zone, the other factors still limit deployments.

There are other sensors that can measure nitrate (but not the other inorganic nutrients) in a moored real-time configuration using by UV absorbance spectrophotometry. For this type of sensor, the limiting factors are: biofouling, power, and precision; it is limited to approximately +/- 2 micromolar NO_3. This relatively low precision may not be to significant in high nitrate environments but its detection limit is above nitrate levels typical in low-nutrient, oligotrophic waters. It is likely that significant progress will continue to be made with both of these technologies and that within a few years, 6 month deployments will become standard.

A lot of progress has been made with real-time moored sensors that are capable of detecting biological organisms including algal species and bacteria through RNA techniques. At present the samplers are large and expensive, and rely on enzymes that carry limitations similar to the reagent of the nutrient sensors; they are bulky and are potent for a limited time. A new approach in currently under development that is

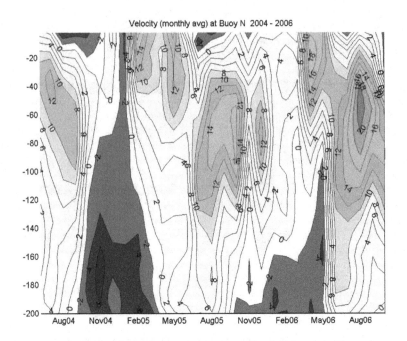

Velocity (monthly avg) at Buoy N 2004 - 2006

Fig. 9. Observed flow through the Northeast Channel. Dark grey shading shows the anomalous deep outflow flow conditions that prevailed from August-April 2004 and September-May 2006. Previous measurements have shown consistent inflow (light grey) in this region with maximum inflows generally occuring in the August to September time period of each year. Flow values presented are montly average values and are expressed in units of cm/sec.

not dependent upon enzymatic action, and once this technique is fully developed longer deployments and orders of magnitude more samples should be possible. These technologies are on the five year horizon and should revolutionize the monitoring of marine microorganisms.

Moored profilers offer tremendous advantages over fixed-point sensor deployments. The most obvious advantage is several orders of magnitude increase in the vertical resolution of the measurements. The ability to profile vertically reduces drastically the number of sensors necessary to effectively monitor the bulk properties of the water column, and it also makes possible the detection of thin layers of high gradient that may be crucial to proper understanding of physical and biogeochemical dynamics.

While several successful profilers are commercially available at this time, they are of the "wire crawler" or bottom winch varieties. These technologies are very successful but they are not ideally suited to the buoy configurations presently used in Go-MOOS and most other observing systems. The wire crawlers require a subsurface taut-wire mooring to successfully climb in the presence of high wave action. Near surface measurements are problematic with such a system as is telemetry. The bottom winch style of profiler need to be protected in heavily fished waters by a group of relatively cumbersome guard buoys.

In order to work with a surface following buoy system like the GoMOOS array, a profiler would need to be decoupled from the large amplitude motions of the buoy and attached wire (10-15 m in winter storm conditions) while still being able to telemeter its data in real time. One possible solution would be a buoyancy-driven profiler. Such a system would rise and descend at a relatively constant rate despite the extreme heaving of the wire under winter storm wave conditions. The data would be transferred to the buoy system by inductively coupling the profiler's sensors to the mooring cable as is presently done with the fixed sensors on the GoMOOS buoys. A "bumper" would stop the profiler one meter below the buoy to avoid damage, and the profiler would then sink back to the near-bottom until the next hour.

Autonomous and semi-autonomous mobile sensor platforms are on the verge of revolutionizing oceanographic surveys. In limited ways these platforms can stand in either ships or moored buoys, and they are generally very cost effective. For the purposes of this discussion we separate these platforms into Autonmous Underwater Vehicles (AUVs) and Autonomous Surface Vehicles (ASVs). The former have advanced much farther in development, although the potential of the ASVs may ultimately be greater.

AUVs are conveniently divided into propeller-driven vehicles that act much like miniature submarines, and gliders that use buoyancy adjustments and wings for locomotion. Both classes of vehicles need to surface in order to transfer their data, receive any changes or adjustments in mission planning, and in most cases, to obtain surface position data for subsurface dead reckoning.

AUVs with active propulsion systems have the advantage of much greater speed, greater maneuverability, and can thus carry out complex quasi-synoptic missions of limited duration. The principal limitation at the present time is mission length. The sensors systems, propulsion systems, and data telemetry/navigation systems are all battery powered. Under the constraints of existing battery technology the submersibles are limited to a few days of operation.

Gliders are much more energy efficient than AUVs that have active propulsion systems, and the long-range seaglider can achieve mission lengths of approximately six months with a range that can take it across an ocean basin while "seesawing" between the surface and 1000m depth. The glider adjusts its buoyancy and shifts its batteries for and aft to achieve nose down sinking and nose up rising. Once the buoyancy and trim are adjusted, no further energy is needed for propulsion until the glider reaches the limit of its descent or ascent and readjusts for the opposite cycle. Since most of the energy usage occurs during the buoyancy adjustments and the data transfer, the mission length is a function of the dive depth. Since shallow water requires more frequent transitions the mission lengths are generally limited to a few weeks.

Gliders and active propulsion AUVs are usually on the order of 2 m in length and 20-30 cm in diameter. Thus they have limited payload capacity that must be shared by sensors and battery packs. As sensor miniaturization evolves we may anticipate that more sensors can be accommodated in these AUVs. In addition, as battery technology improves, mission range and duration should become less serious constraints. In addition to the obvious role in substituting for extremely expensive manned surveys, AUVs (particularly gliders) may be able to serve as short-term virtual moorings. By gliding back and forth, the glider is able to profile at essentially one location (assuming the horizontal scales of interest are sufficiently large) to obtain a time serious of profiles.

Within the next five years we expect to see a glider that is capable of landing on the bottom and making a set of measurements (including acoustic Doppler current profiles) while stationary, then making a profile of water properties (including temperature, salinity, dissolved oxygen, and chlorophyll fluorescence) while rising vertically to the surface and then gliding to the next "moored" location. This glider/lander would act much like an array of fixed moorings.

Although ocean surface drifters have been in scientific use for at least as far back as the Challenger Expedition in the 1870's, their usefulness was limited by the inability to track them over long distances with more resolution than the release and recovery points. This technical problem was alleviated a hundred years later in the 1970's when the ARGOS satellite system was developed that allowed positions of a large number of drifters fixed multiple times per day over a broad area. This development ushered in the modern era of satellite-tracked drifters. Since the advent of Global Positioning System (GPS), it has been possible to track drifters continuously and to telemeter these data via global and local communications including those built into the NOAA's Geostationary Operational Environmental Satellites (GOES) weather satellite system.

Coastal Ocean Observing Systems (COOS) may benefit from the development of a system tailored to the scales of coastal environments. The density of drifters mapping the small and meso surface circulation and frontal structures needs to be very high. Conversely, the range of the built-in telemetry systems need not be global, and the deployments may be of shorter duration. We foresee the development and application of inexpensive wireless drifters that form a self-organizing network of relayed, multi-path communications between neighbors. Individual drifters would broadcast their positions and ancillary data streams (such as temperature, wave heights, etc.) as well as the data of all neighbors within range. In this way, a few inexpensive drifters within range of the array of nodes would transfer the data of the entire fleet to the observing system. The relatively inexpensive GPS drifters could be deployed in large numbers in order to resolve meso-scale features, directly measure vorticity, and estimate dispersion. A larger-scale buoy array with long range communication capabilities (such as GoMOOS) could serve as a grid of communication nodes. In cases where the associated buoy array is widely spaced, a floating node could be deployed with the drifters in order to maximize the number of drifters that remain within network. Air-deployable nodes dropped into the centroid of the drifter distribution could be deployed as the drifters begin to disperse widely. Another possibility is that autonomous or remote-controlled aerial drones may act as data mules to recover the data from drifters that have strayed from the network.

4 Summary

The GoMOOS system is a prototype for coastal observing systems in the United States. It has been remarkably successful in its first years of operation, and has proven the feasibility of the concept. The real-time data are now routinely used by mariners, fisherman, recreational boaters, aquaculturists, marine pilots, the US Coast

Guard, the national weather service, natural resource managers, oceanographers, and the general public. In this regard, it is fair to say that the system has exceeded expectations.

In addition, the wealth of time series data has already begun to yield scientific results that would not have been possible without the long-term records that are unprecedented in this region. We anticipate that the scientific contributions will increase greatly as the data records become longer. Continued progress in sensor development can be expected to expand greatly the biological and chemical time-series measurements that will become feasible in near future.

Acknowledgments

We thank the entire staff of the Physical Oceanography Group of the University of Maine and the Optics Group at Bigelow Laboratory, whose dedication and determination made this work a success. We gratefully acknowledge funding from ONR, NOAA, the State of Maine, the University of New Hampshire, and the University of Maine, and we also acknowledge the efforts of the GoMOOS staff and the GoMOOS Board of Directors who have worked to make these data accessible to the public via www.gomoos.org.

References

1. Xue, H., Shei, L., Cousins, S., Pettigrew, N.R.: The GoMOOS nowcast/forecast system. Cont. Shelf Res. 25, 2122–2146 (2005)
2. Pettigrew, N.R., Townsend, D.W., Wallinga, J.P., Brickley, P.J., Hetland, R.D., Xue, H.: Observations of the Eastern Maine Coastal Current and its Offshore Extensions in 1994. Journal of Geophysical Research 103(C13), 30,623–30,639 (1998)
3. Sinclair, M., Wilson, S., Subba Rao, D.V.: Overview of the biological oceanography of the Gulf of Maine. In: Proceedings of the Gulf of Maine Scientific Workshop, Woods Hole Massachusetts, pp. 91–111 (1991)
4. Bigelow, H.B.: Physical Oceanography of the Gulf of Maine. Fish. Bull. 511-1027 (1927)
5. Brooks, D.A.: Vernal circulation in the Gulf of Maine. Jour. of Geophys. Res. 90, 4687–4705 (1985)
6. Pettigrew, N.R., Churchill, J.H., Janzen, C.D., Mangum, L.J., Signell, R.P., Thomas, A.C., Townsend, D.W., Wallinga, J.P., Xue, H.: The kinematic and hydrographic structure of the Gulf of Maine Coastal Current. Deep Sea Res. II 52, 2369–2391 (2005)
7. Wallinga, J.P., Pettigrew, N.R., Irish, J.D.: The GoMOOS moored Buoy Design. In: Proceedings of the IEEE Oceans 2003 conference, pp. 2596–2599 (2003)
8. Siegel, D.A., et al.: Bio-optical modeling of primary production on regional scales: The Bermuda BioOptics project. Deep-Sea Res. II 48, 1865–1896 (2001)
9. Roesler, C.S., Boss, E.: In situ measurements of the inherent optical properties (IOPS) and potential for harmful algal bloom (HAB) detection and coastal ecosystem observations. ch. 5. In: Babin, M., Cullen, J.J., Roesler, C.S. (eds.) Real-time Coastal Observing Systems for Ecosystem Dynamics and Harmful Algal Blooms. UNESCO Series Monographs on Oceanographic Methodology (2005)

10. Pettigrew, N.R., Roesler, C.S.: Implementing the Gulf of Maine Ocean Observing System (GoMOOS). In: IEEE proceedings of Oceans 2005 (Europe), Brest France, pp. 1234–1241 (2005)
11. Roesler, C.S., Boss, E.: Spectral bean attenuation coefficient retrieved from ocean color inversion. Geophys. Res. Lett. 30, 1468–1472 (2003)
12. Pettigrew, N.R., Wallinga, J.P., Neville, F.P., Schlenker, K.R.: Gulf of Maine Ocean Observing System: Current Measurement Approaches in a Prototype Integrated Ocean Observing System. In: IEEE Proceedings, Eighth Current Measurement Technology Conference 2005, pp. 127–131 (2005)

Using the Sensor Web to Detect and Monitor the Spread of Vegetation Fires in Southern Africa

Andrew Terhorst[1], Deshendran Moodley[2], Ingo Simonis[3], Philip Frost[1],
Graeme McFerren[1], Stacey Roos[1], and Frans van den Bergh[1]

[1] Meraka Institute, CSIR, PO Box 395, Pretoria, 0001, South Africa
{aterhorst,pfrost,gmcferren,sroos,fvdbergh}@meraka.org.za
http://ict4eo.meraka.org.za
[2] School of Computer Science, University of Kwazulu Natal, Durban,
4041, South Africa
moodleyd37@ukzn.ac.za
http://www.sciag.ukzn.ac.za/comp/
[3] Margarete-Bieber Weg 11, 35396, Giessen, Germany
ingo.simonis@geospatialresearch.de
http://www.geospatialresearch.de

Abstract. Key concepts in disaster response are level of preparedness, response times, sustaining the response and coordinating the response. Effective disaster response requires a well-developed command and control framework that promotes the flow of information. The Sensor Web is an emerging technology concept that can enhance the tempo of disaster response. We describe how a satellite-based system for regional vegetation fire detection is being evolved into a fully-fledged Sensor Web application.

1   Introduction

Most disasters are of short duration and require a fixed amount of consequence management. Examples include earthquakes, tsunamis and storm events. Other disasters are more complex and unfold in a non-linear fashion over an extended period. Such disasters require ongoing and adaptive consequence management. Examples include the outbreak contagious diseases (*e.g.* bird flu) and wild fires.

Key concepts in disaster response are level of preparedness, response times, sustaining the response and coordinating the response [1]. Time is critical and the three primary challenges in the race against time are *uncertainty, complexity* and *variability* [2]. Dealing with these challenges requires a well-designed command and control framework that promotes the free flow of information. Many consider the foundation for command and control to be the Observe-Orient-Decide-Act (OODA) loop [3]. The time it takes to complete an OODA cycle is what determines the tempo of the disaster response (Figure 1).

S. Nittel, A. Labrinidis, and A. Stefanidis (Eds.): GSN 2006, LNCS 4540, pp. 239–251, 2008.

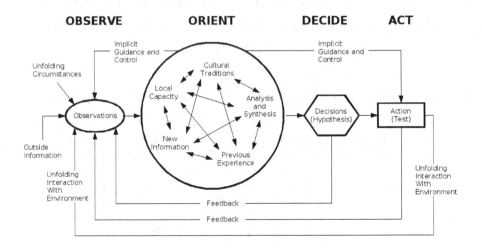

Fig. 1. The OODA loop

Advances in sensor technology and distributed computing, coupled with the development of open standards that facilitate sensor/sensor network interoperability, are contributing to the emergence of a phenomenon known as the 'Sensor Web'[4]. This phenomenon can be described as an advanced Spatial Data Infrastructure (SDI) in which different sensors and sensor networks are combined to create a sensor-rich feedback control paradigm [5]. In this paper, we describe how the Sensor Web can enhance the tempo of disaster response in context of vegetation fires.

2   Sensor Web Enablement

Sensor Web Enablement (SWE) is an Open Geospatial Consortium (OGC) initiative that extends the OGC open web services framework [6] by providing additional services for integrating web-connected sensors and sensor systems. SWE services are designed to enable *discovery* of sensor assets and capabilities, *access* to these resources through data retrieval and subscription to alerts, and *tasking* of sensors to control observations [7]. SWE is not restricted to sensors/sensor systems but also refers to associated observation archives, simulation models and processing algorithms. SWE enables interoperability between disparate sensors, simulation models and decision support systems. It acts as a middleware layer that connects physical assets, geo-processing applications and decision support tools. SWE provides services that automatically trigger actuators when certain conditions are true. The following figure illustrates the SDI and the advanced SWE model.

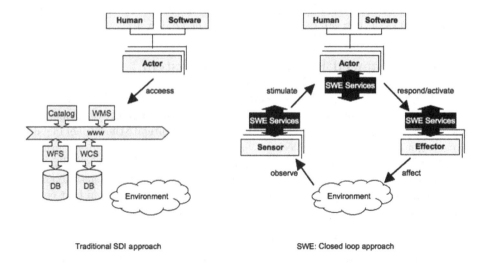

Fig. 2. OGC Compliant AFIS Architecture

2.1 SWE Information Model

The SWE initiative has developed draft specifications for modelling sensors and sensor systems (SensorML, TransducerML), observations from such systems (Observations and Measurements) and processing chains to process observations (SensorML) [8,9]. The draft specifications provide semantics for constructing machine-readable descriptions of data, encodings and values, and are designed to improve prospects for plug and play sensors, data fusion, common data processing engines, automated discovery of sensors, and utilisation of sensor data.

2.2 SWE Services Model

SWE provides four types of web services: Sensor Observation Service (SOS), Sensor Alert Service (SAS), Sensor Planning Service (SPS) and Web Notification Service (WNS) [10,11,12,13]. The SOS provides a standard interface that allows users to retrieve raw or processed observations from different sensors, sensor systems and observation archives. The SAS provides a mechanism for posting raw or processed observations from sensors, process chains or other data providers (including a SOS) based on user-specified alert/filter conditions. When subscribing to a SAS, users not only define the alert conditions but also the communication protocol for disseminating alerts via the WNS.

The WNS provides a standard interface to allow asynchronous communication between users and services and between different services. A WNS is typically used to receive messages from a SAS and to send/receive messages to and from a SPS. The SPS provides a standard interface to sensors and sensor systems and is used to coordinate the collection, processing, archiving and distribution of sensor observations. Discovery of OGC and SWE services is facilitated by the Sensor Web Registry Service – an extended version of the OGC Catalogue Service [14].

3   The Advanced Fire Information System

The Advanced Fire Information System (AFIS) is the first near real-time satellite-based fire monitoring system in Africa. It was originally developed for the South African electrical power utility, ESKOM, to mitigate the impact of wild (vegetation) fires on regional electricity supply [15,16]. Vegetation fires burning under high-voltage transmission lines can lead to line faults (flash-overs) disrupting regional electricity supply.The destruction caused by seasonal vegetation fires emphasised the need to develop a satellite-based information system that could provide information on the frequency and distribution of vegetation fires to the research and fire fighting community.

ESKOM and the CSIR commenced research to investigate the efficacy of satellite thermal infrared sensors to detect vegetation fires that could cause flash-overs anywhere along the 28000 km of transmission lines that criss-cross South Africa. AFIS, implemented in June 2004, searches for hotspots within a 5km buffer zone along all transmission lines at 15 minute time intervals. When a hotspot is detected, AFIS generates email and SMS text messages that are transmitted to relevant authorities in near real-time. AFIS was first implemented using propriety GIS technology developed by the University of Maryland (Web Fire Mapper,[17]) but has now been re-engineered as an OGC compliant Sensor Web application based on open source software.

3.1   Hotspot Detection

AFIS currently relies on a contextual algorithm for hotspot detection using the MODIS sensor aboard the polar orbiting TERRA and AQUA satellites and the SEVIRI sensor aboard the geostationary METEOSAT-8 satellite [18]. The hotspot update rate for MODIS is every six hours compared to every 15 minutes for SEVIRI. Though the SEVIRI provides almost near real-time hotspot detection, it can only resolve hotspots five hectares or more in extent unlike MODIS, which can resolve hotspots less than a hectare in size.

The hotspot detection algorithm was originally developed for the Advanced Very High Resolution Radiometer (AVHRR) sensor flown aboard the TIROS satellites [18,19,20]. The algorithm uses 3.9μm and 10.8μm bands to discriminate fire pixels from background pixels. The algorithm first classifies a pixel according to a fixed threshold, e.g. $T > 300K$, to identify potential fire pixels – the remaining pixels are called *background* pixels. The neighbourhood of this pixel is then searched for background pixels, growing the neighbourhood if necessary to ensure that at least 25% of the neighbourhood pixels are background pixels. From this set of background pixels, the mean and standard deviation statistics are calculated from the 3.9μm and the 3.9μm $- 10.8\mu$m band difference data. The pixel under consideration is then classified as a hotspot if its

3.9μm value exceeds the background mean by some multiple of the standard deviation – a similar test is performed on the 3.9μm – 10.8μm band difference.

3.2   Current Sensor Web Architecture

Figure 3 depicts the current OGC compliant architecture for AFIS. The contextual algorithm for hotspot detection has been implemented in two separate image processing chains - one dedicated to MODIS imagery, the other dedicated to SEVIRI imagery. The image processing chains record hotspot events in a database that is exposed to the Fire Alert Service (FAS) via a SOS. The FAS essentially is a SAS with some additional application logic for spatial processing. Spatial processing is currently limited to intersecting hotspot events with features of interest. However, the aim is to enrich fire alerts by populating hotspot events with additional attribute data such as surface wind vectors and fire danger index provided by other OGC web services such as a Web Coverage Service (WCS) and/or Web Feature Service (WFS) and/or another SOS.

The FAS has been implemented within the OXFramework, a Graphical User Interface (GUI) developed by 52°North Open Source Spatial Data Infrastructure Initiative (www.52North.org). When subscribing to the FAS, users must specify the preferred medium for receiving fire alerts (*e.g.* Simple Message Service (SMS) or Email) and what parameters to pass through to the spatial process chain. Parameters may include what features of interest to intersect with hotspots (in the case of ESKOM, this would be buffer zones around high-voltage transmission lines).

4   Results and Discussion

4.1   Hotspot Detection Success Rate

The success of AFIS as a management tool within ESKOM is measured by its ability to provide early detection of hotspots (fires) close to transmission lines before flash-overs occur. MODIS was able to detect an average of 44% of all flash-over fires between 2003 and 2005 whereas SEVIRI detected 46% of all flash-over fires during the same period. By combining the detection accuracy of MODIS and MSG within one system (AFIS), the detection accuracy rose to 60%. The statistics of the MODIS and SEVIRI detections clearly demonstrate the limitations of these current sensors as a detection tool on their own. The MODIS sensor was able to detect many of the smaller fires, but due to its infrequent revisit time was unable to detect more than 40% of the fires. The SEVIRI sensor on the other hand struggled to detect smaller fires due to its coarse resolution. The 2% higher detection accuracy calculated for SEVIRI with its lower resolution and less advanced detection algorithm shows the importance of frequent observations. The combination of accurate detection from MODIS with the frequent detection from SEVIRI increased the fire detection rate by more than 15%[16].

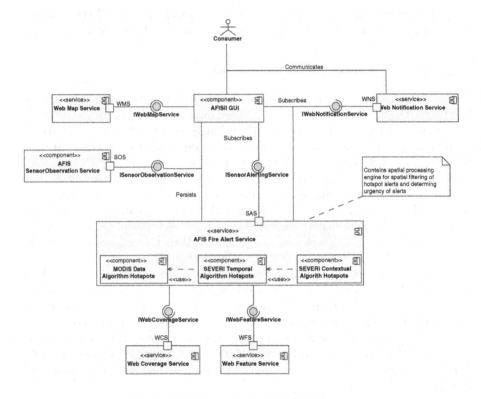

Fig. 3. OGC Compliant AFIS Architecture

To improve the detection rate, a new non-contextual hotspot algorithm for the SEVIRI sensor that is more sensitive is being developed. The basic approach is to build a general model of the diurnal cycle for the thermal infrared band, and then to fit this model to the observed data of the last 24 hours. This model can then be used to generate accurate estimates of the expected background temperatures. If a statistically significant difference between the current observed temperature and the predicted background temperature is observed, then the pixel in question is classified as a hotspot. The first implementation of this algorithm relied on a Kalman filter to provide the estimates of the background temperature. Initial results indicate that this method is significantly more sensitive, particularly in cases where the background temperature is below 300K (*e.g.* as in the early morning hours)[21].

4.2   Extending AFIS Functionality

The intention is to shift the emphasis from simple hotspot detection to more sophisticated fire risk management. This requires a good understanding of what controls vegetation fire behaviour. We are currently building a domain ontology

[22,23] for vegetation fires. The ontology will capture key concepts in the vegetation fire domain such as combustion properties, fuel load, burning regime, fire weather, fire suppression methods, and topographical controls. The aim is to use the Sensor Web to observe specific fire-related phenomena described in the vegetation fire ontology and employ machine reasoning to determine fire risk *i.e.* automate the *observe* and *orient* parts of the OODA loop and issue more meaningful fire alerts.

4.3   Sensor Web Agent Platform

To achieve the desired level of automation requires a more intelligent architecture than what the current SWE framework provides. We are advocating an open, service-oriented, multi-agent system architecture for the Sensor Web known as the Sensor Web Agent Platform (SWAP) [24]. This architecture is a hybrid of the Foundation for Intelligent Physical Agents (FIPA)[25] and OGC standard architectures. SWAP incorporates the following concepts: Ontologies, process models, choreography, service directories and facilitators, service-level agreements and quality of service measures [26]. Ontologies will provide explicit descriptions of components within SWAP *i.e.* sensors and sensor data, simulation models, algorithms and applications, and how these components can be integrated and used by software agents.

Ultimately, users should be able to improve or alter the behaviour of SWAP by editing the underlying ontologies. SWAP uses the Web Ontology Language (OWL) as the ontology representation language [27]. The SWAP ontology set is split along the three cognitive dimensions of space, time and theme [28], with a fourth dimension for representing data structures (Figure 4). It consists of a *swap-theme* ontology that contains thematic concepts, the *swap-space* ontology that contains spatial concepts, the *swap-time* ontology that contains temporal concepts, and the *swap-data* ontology that contains concepts for representing data structures. These *upper* ontologies [29] provide an extensible framework that grounds all other ontologies in SWAP. Domain ontologies, *i.e.* ontologies for specific application domains, are built by extending the *swap-theme* ontology. For example, the vegetation fire ontology contains concepts for building applications in the fire risk management domain and uses concepts from an existing earth science ontology, the NASA SWEET ontology [30]. These ontologies represent an *online model* of the system, such that any changes in the ontology are dynamically integrated into the system.

The vegetation fire ontology is being constructed with several objectives in mind. Primarily, it serves to inform SWAP agents about the things that need to be reasoned about when fulfilling the AFIS use cases of detecting vegetation fires and predicting their behaviour to assess risk. Specifically, the ontology describes the observational properties of concepts and relationships in the vegetation fire domain. In other words, what is it that allows a certain concept to be observed? The vision is that agents in the SWAP will marry observational properties to appropriate web-accessible sensors or sensor services appropriate for gathering the desired observation. We take a similar view to Pennington [31] that observation

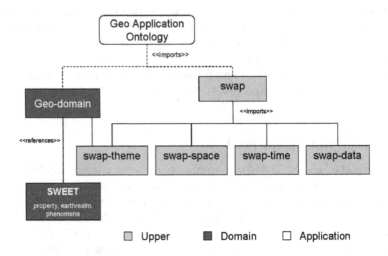

Fig. 4. SWAP Ontology Architecture

links the high level categories of theory, space, time and entitities in order that process can be understood. We thus concern ourselves with modelling knowledge about the spatial-temporal aspects of fire (spread, intensity), the entities upon which fire depends (fuel, topography, weather) and theories or understandings of how they interact. Observation of the attributes of these ontological categories allows us to populate process models with values, leading us toward answers to questions such as "in which direction will this fire spread?" or "is this fire a danger to this infrastructure?". Users can improve or alter the behaviour of SWAP by editing any of the underlying ontologies.

The SWAP abstract architecture is split into three layers: *Sensor Layer*, *Knowledge Layer* and *Application Layer* (Figure 5). The Sensor Layer is populated by sensor agents that encapsulate individual sensors, sensor systems and archived observations. They expose sensor data in a uniform way and deal with any sensor-dependant processing. Data from sensor agents form input to agents in the second, Knowledge Layer, which consist of workflow, tool and modeling agents. Workflow agents receive data from sensor agents and pass this data through a combination of tool and modeling agents and aggregate the results. Tool agents provide feature extraction and image processing functionality, for example, while modeling agents encapsulate predictive models and can provide projection and data analysis functionality. The processed data stored by workflow agents forms input to application agents in the Application Layer. Application agents combine higher level features provided by workflow agents and provide different views of these data to different end users. Advanced users would be able to compose and deploy new workflow agents. Typical functionality provided by application agents includes enabling users to specify alert conditions in the system. End users would likely subscribe to more than one application in the system. The user agent allows a user to integrate data from different application

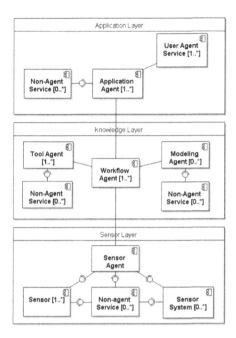

Fig. 5. SWAP Abstract Architecture

agents. Furthermore, since these data are semantically marked up, they can be easily integrated with systems within the user's organisation.

A prototype implementation of SWAP is currently being developed for AFIS (Figure 6). The prototype will use the aforementioned contextual algorithm that compares individual pixel blackbody temperature values to those of its neighbouring pixels, to detect pixel blackbody temperature values beyond a specified threshold. A sensor agent will offer these SEVIRI blackbody temperature values and feed them to a workflow agent configured to detect temperature hotspots. The workflow agent receives these data and tasks the contextual algorithm tool agent to check for hotspots, and communicates detected hotspots to the AFIS application server. The AFIS application server (application agent) uses hotspots retrieved from the workflow agent to issue fire alerts using a SAS and WNS. The AFIS application server may also be accessed via the AFIS client (user agent) installed on a user's computer. The application can be extended by implementing additional agents *e.g.* a fire spread modeler (modeling agent) that retrieves current weather data (*e.g.* wind, humidity and temperature) from a weather SOS via a weather agent. The fire spread modeler will use the weather data, and other internal models like terrain and vegetation cover to predict the spread of detected fires. Another possible extension is to incorporate brightness temperature from the MODIS sensor into the system.

In terms of the OODA loop, the application layer is where orientation and decision making occurs, whereas the knowledge layer facilitates observations by

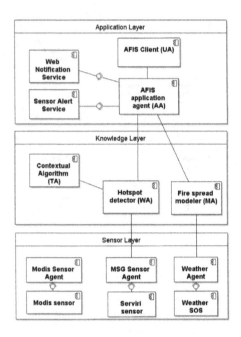

Fig. 6. SWAP Architecture for AFIS Prototype

providing information extraction and data filtering. The knowledge layer can also be extended to coordinate actions, by tasking individual sensors in the sensor layer.

5 Conclusions and Future Work

AFIS is a good example of what the Sensor Web can do to facilitate effective disaster response. The Sensor Web enhances the OODA loop by providing several mechanisms for sensor-rich feedback control. Our proposed service-oriented, multiagent systems architecture for the Sensor Web (SWAP) extends the current SWE framework. SWAP facilitates technical and semantic interoperability and promotes re-usability. The architecture is flexible and extensible: New agents can be deployed or swapped out in a 'plug and play' fashion. Workflows can be easily updated or built from scratch by editing existing or creating new application ontologies. Parts of the OODA loop can be offloaded onto software agents that use machine-reasoning to automatically generate hypotheses. Speeding up the OODA cycle in this way should enhance the tempo of disaster response.

A number of research questions must be addressed before SWAP can be fully realised. These include questions relating to the internal model of agents, communication between agents and between agents and non-agent services, message structure and message payload structure, framework for building ontologies, how to handle contradictory knowledge, how to integrate different types of ontologies into the agent paradigm, maintenance of ontologies, data fusion, dynamic configuration of process chains and appropriate agent development framework.

We intend using the implementation of the SWAP prototype for AFIS to refine these research questions and expose others. Our plan is to work in close collaboration with standards generating bodies such as OGC, FIPA, IEEE and W3C, other research partners and the Open Source Software (OSS) development community to promote SWAP as a reference architecture for the Sensor Web.

Acknowledgements

AFIS is an initiative of the Meraka Institute (part of the Council for Scientific and Industrial Research (CSIR)). ESKOM funded the initial development of AFIS. The re-engineering of AFIS as an OGC compliant platform and research and development of SWAP is being funded by the CSIR. SWAP is the brainchild of a global research partnership known as the Sensor Web Alliance (www.sensorweb-alliance.org).

References

1. Annoni, A., Atkinson, M., Denzer, R., Hecht, L., Millot, M., Pichler, G., Sassen, A.M., Couturier, M., Alegre, C., Sassier, H., Coene, Y., Marchetti, P.G.: Towards an Open Disaster Risk Management Service Architecture for INSPIRE and GMES (version No.9) (February 2005)
2. Rosen, J., Grigg, E., Lanier, J., McGrath, S., Lillibride, S., Sargent, D., Koop, C.E.: The Future of Command and Control for Disaster Response. IEEE Engineering in Medicine and Biology 21(5), 56–68 (2002)
3. Osinga, F.: Science, Strategy and War: The Strategic Theory of John Boyd. Eburon Academic Publishers, Delft (2005)
4. Liang, S.H.L.C., Tao, C.V.: A Distributed Geospatial Infrastructure for the Sensor Web. Computers & Geosciences 31(2), 221–231 (2005)
5. Zibikowski, R.: Sensor-Rich Feedback Control: A New Paradigm for Flight Control Inspired by Insect Agility. Instrumentation and Measurement Magazine 7(3), 19–26 (2004)
6. OGC: The OGC Abstract Specification - Topic 0: Abstract Specification Overview. Document 04-084, Open Geospatial Consortium, Wayland, Massachussets, USA (2005) (accessed April 14, 2006),
   http://portal.opengeospatial.org/files/?artifact_id=7560
7. Botts, M., Robin, A., Davidson, J., Simonis, I.: OpenGIS Sensor Web Enablement Architecture Document. Document 06-021r1, Wayland, Massachussets, USA (March 2006) (accessed July 4, 2006),
   http://portal.opengeospatial.org/files/?artifact_id=14140
8. Botts, M.: OpenGIS Sensor Model Language (SensorML). Document 05-086, Wayland, Massachussets, USA (November 2005) (accessed Febuary 2006),
   http://portal.opengeospatial.org/files/?artifact_id=12606
9. Cox, S.: Observations and Measurements. Document 05-087r5, Wayland, Massachussets, USA (April 2005) (accessed May 5, 2006),
   http://portal.opengeospatial.org/files/?artifact_id=14034

10. Simonis, I., Wytzisk, A.: Web Notification Service. Document 03-008r2, Wayland, Massachussets, USA (April 2003) (accessed Febuary 6, 2006), http://portal.opengeospatial.org/files/?artifact_id=1367
11. Na, A., Priest, M.: Sensor Observation Service. Document 05-088r1, Wayland, Massachussets, USA (January 2006) (accessed April 1, 2006), http://portal.opengeospatial.org/files/?artifact_id=12846
12. Simonis, I.: Sensor Planning Service. Document 05-089r1, Wayland, Massachussets, USA (December 2005) (accessed April 1, 2006), http://portal.opengeospatial.org/files/?artifact_id=12971
13. Simonis, I.: Sensor Alert Service. Document 06-028, Wayland, Massachussets, USA (April 2006) (accessed July 4, 2006), http://portal.opengeospatial.org/files/?artifact_id=13921
14. Nebert, D.: OpenGIS Catalogue Service Implementation Specification. Document 04-021r3, Wayland, Massachussets, USA (August 2004) (accessed, September 2005), http://portal.opengeospatial.org/files/?artifact_id=5929&version=2
15. Fleming, G., van den Bergh, F., Claudel, F., Frost, P.: Sensor Web Enabling the Advanced Fire Information System. In: Proceedings of the 2005 International Symposium for Environmental Software Systems (ISESS 2005), Hotel do Mar, Sesimbra, Portugal (2005) (accessed July 4, 2006), http://www.isess.org/documents/2005/presentations/
16. Frost, P., Vosloo, H.: Providing Satellite-Based Early Warnings of Fires to Reduce Fire Flashovers on South African Transmission Lines. In: Proceedings of the 10th Biennial Australasian Bushfire Conference: Bushfire 2006, ICMS (June 2006), http://www.fireandbiodiversity.org.au/
17. Davies, D., Kumar, S., Descloitres, J.: Global Fire Monitoring: Use of MODIS Near-real-time Satellite Data. Geomatics Info Magazine 18(4) (2004)
18. CEOS: Earth Observation Handbook. Committee on Earth Observation Satellites (2002), http://www.ceos.org/pages/pub.html#handbook
19. Flasse, S.P., Ceccato, P.: A Contextual Algorithm for AVHRR Fire Detection. International Journal of Remote Sensing 17(2), 419–424 (1996)
20. Giglio, L., Descloitres, J., Justice, C., Kaufmann, J.: An Enhanced Contextual Fire Detection Algorithm for MODIS. Remote Sensing of Environment 87, 273–282 (2003)
21. van den Bergh, F., Frost, P.: A Multi-Temporal Approach to Fire Detection using MSG Data. In: Proceedings of Multitemp 2005, Biloxi, Mississippi, USA (2005)
22. Gruber, T.R.: Towards Principles for the Design of Ontologies Used for Knowledge Sharing. Kluwer Academic Publishers, Deventer (1993), http://citeseer.ist.psu.edu/gruber93toward.html
23. Struder, R., Benjamins, V.R., Fensel, D.: Knowledge Engineering: Principles and Methods. IEEE Transactions on Data and Knowledge Engineering 25(1), 161–197 (1998)
24. Moodley, D., Simonis, I.: A New Architecture for the Sensor Web: the SWAP-Framework. In: Cruz, I., Decker, S., Allemang, D., Preist, C., Schwabe, D., Mika, P., Uschold, M., Aroyo, L.M. (eds.) ISWC 2006. LNCS, vol. 4273. Springer, Heidelberg (2006)
25. FIPA: FIPA Abstract Architecture Specification. Technical Specification SC00001L, Foundation for Intelligent Physical Agents, Geneva, Switzerland (2002), http://www.fipa.org

26. Huhns, M., Singh, M., Burstein, M., Decker, K., Durfee, K., Finin, T., Gasser, T., Goradia, H., Jennings, P., Lakkaraju, K., Nakashima, H., Van Dyke Parunak, H., Rosenschein, J., Ruvinsky, A., Sukthankar, G., Swarup, S., Sycara, K., Tambe, M., Wagner, T., Zavafa, L.: Research Directions for Service-Oriented Multiagent Systems. IEEE Internet Computing, 65–70 (November 2005)

27. Bechhofer, S., van Harmelen, F., Hendler, J., Horrocks, I., McGuinness, D.L., Patel-Schneider, P.F., Stein, L.A.: OWL Web Ontology Language Reference (February 2004) (accessed July 11, 2006),
http://www.w3.org/TR/2004/REC-owl-ref-20040210/

28. Mennis, J.L., Peuquet, D.J., Qian, L.: A conceptual framework for incorporating cognitive principles into geographical database representation. International Journal of Geographical Information Science 14(6), 501–520 (2000)

29. Guarino, N.: Formal Ontology in Information Systems. IOS Press, Amsterdam (1998)

30. Raskin, R.: Guide to SWEET Ontologies for Earth System Science. Technical report (January 2006) (accessed, February 2006),
http://sweet.jpl.nasa.gov/guide.doc

31. Pennington, D.: Representing the Dimensions of an Ecological Niche. In: Cruz, I., Decker, S., Allemang, D., Preist, C., Schwabe, D., Mika, P., Uschold, M., Aroyo, L.M. (eds.) ISWC 2006. LNCS, vol. 4273. Springer, Heidelberg (2006)

Peer-to-Peer Shared Ride Systems

Yun Hui Wu, Lin Jie Guan, and Stephan Winter

Department of Geomatics, The University of Melbourne, Victoria 3010, Australia

Abstract. Shared ride systems match the travel demand of transport clients with the supply by vehicles, or hosts, such that the clients find rides to their destinations. A peer-to-peer shared ride system allows clients to find rides in an ad-hoc manner, by negotiating directly with nearby hosts via radio-based communication. Such a peer-to-peer shared ride system has to deal with various types of hosts, such as private cars and mass transit vehicles. Their different behaviors affect the negotiation process, and consequently the travel choices. In this paper, we present and discuss a model of a peer-to-peer shared ride system with different types of agents. The behavior of the model is investigated in a simulation of different communication and way-finding strategies. We demonstrate that different types of agents enrich the choices of the clients, and lead to local solutions that are nearly optimal.

1 Introduction

Research on geosensor networks is typically concerned with the efficient extraction of information of sensor observations, hence, looking into hardware, protocols, routing of messages, and data aggregation. The research presented in this chapter is different in some respects. First of all, its focus lies on movement of the nodes, not on movement of information. The investigated geosensor network consists of nodes that have individual, specific travel intentions. If two nodes meet, one of them can ride piggy-back on the other one for reasons like saving fuel or traveling faster, depending on the abilities of the two nodes. Secondly, this geosensor network allows for different classes of nodes. In applications, one will distinguish transportation clients from transportation hosts. Furthermore, different clients and hosts can be distinguished. For example, there may be clients that can move only with a host, otherwise they are static, or there may be clients that travel significantly slower than hosts. Finally, this geosensor network underlies the well-known communication constraints of all geosensor networks. Nodes have to communicate to match clients with hosts, but communication is limited to local neighborhoods because of scarce resources in terms of battery and bandwidth, and because of a fragile communication network topology due to node mobility.

The interesting research questions in this context are about communication and trip planning strategies of nodes, about global optimization of trips from local transportation network knowledge, and about the general behavior of large transportation geosensor networks with autonomous nodes. This chapter will address and illuminate the questions by a concrete realization: a shared ride system for persons traveling by multiple modes in the city.

Movement of people in a city forms a complex system. It includes the street network and other ways of traveling, traffic rules, traffic infrastructure (e.g., traffic lights,

S. Nittel, A. Labrinidis, and A. Stefanidis (Eds.): GSN 2006, LNCS 4540, pp. 252–270, 2008.

signs) as well as cognition, decisions and actions of intelligent, autonomous agents such as pedestrians and vehicle drivers. This complex system is burdened by more and more traffic in expanding cities. In this situation a peer-to-peer shared ride system can provide relief to the critical situation: it enables people to negotiate in an ad-hoc manner for ride sharing, and thus, helps reducing traffic, increases urban accessibility, and improves the integration of different modes of transport. In such a system, pedestrians are the agents with transport demand, called clients, and vehicles, or hosts, provide the transport supply. Finding rides in an ad-hoc manner is accomplished by local negotiation between these agents via radio-based communication.

A peer-to-peer shared ride system has to deal with various types of agents, such as private cars and mass transit vehicles, or mobile and immobile clients, to cope adequately with the complexity of urban movements. The agents' different interests, capacities and behaviors affect the negotiation process, and consequently, the trips undertaken. For example, hosts can be distinguished by their travel speed, their passenger capacity and their fare structure, and clients can be distinguished by their mobility.

In this situation a client cannot stay with a simple preference for one mode of traveling, that is, one type of hosts. For example, in general a rushed client would prefer hosts can deliver a quick and direct trip: taxis. On the other hand, taxis can be in high demand during peak travel times and catching trams, trains or buses can be an alternative: they may travel slower but might reach the destination earlier depending on traffic. Hence, in this paper we present and discuss a model of a peer-to-peer shared ride system with different types of agents.

Agents, that is, clients and hosts in peer-to-peer shared-ride systems have knowledge of their environment. They can collect and transmit information from/to their neighbors. Frequently agents have choices. They have preferences, various optimization criteria, such as money or time, and are able to make current optimal decisions based on their knowledge. However, for practical reasons agents have only local and current knowledge of their environment. Previous research [1] investigates the ability to make trip plans from different levels of local knowledge. It shows that a mid-range communication depth is both efficient (needing less communication messages than complete current knowledge) and effective (leading to a travel time comparable to complete current knowledge). This investigation was based on a simulation with homogeneous hosts and an immobile client. This paper poses the hypothesis that involving other types of agents, the trips will change significantly, but mid-range communication is still both efficient and effective compared to other communication strategies.

This hypothesis will be approached by simulation. The simulation is realized as a multi-agent system, allowing us to model and understand individual behavior of different agents. The approach requires identifying and specifying the essential aspects of an urban shared ride system, implementing them in a multi-agent system, and then running large numbers of random experiments to generate the required evidence. The model can be investigated by systematically varying the design parameters and studying the peer-to-peer shared ride system behavior.

This paper has the following structure. Section 2 reviews previous and related research. Section 3 discusses the types of agents in shared ride systems. Section 5 presents the design of a multi-agent simulation, and the simulation results are provided in Section 6. Section 7 concludes with a discussion and future work.

2   Literature Review

This review consists of a literature overview of shared ride systems in general, and of agent-based simulation of shared ride system in particular, with special attention to previous research on a peer-to-peer shared ride system.

2.1   Shared Ride Systems

In the real world, shared ride systems exist in many forms and names, such as *carpooling*, *vanpooling*, *dial-a-ride*, or *find-a-ride*. Shared ride systems also have various levels of technological support, such as being based simply on social convention, or using a centralized database with pre-registration and/or pre-booking via a Web interface.

Carpooling/Vanpooling can be seen as a prearranged shared ride service between home and workplace to save up parking spaces [2]. Traditional carpooling/vanpooling services are organized by private companies and are not door-to-door. People with regular commuting schedules usually meet in a place to share vehicles running on prearranged times and routes. Van pooling is limited by the provider's service area and not viable for areas or individual origins or destinations that do not have the critical mass of people using the service. New users can only participate in existing poolings, or they can create a new pooling with others.

Mass transit systems, like the underground, trains, buses and trams, run on predefined schedules and routes. Being government funded or subsidized, the fares are typically lower than the costs of private means of transportation. In addition to guaranteeing mobility and access for everybody, this shall also encourage people to mitigate individual car traffic. However, such a shared ride is restricted to fixed time schedules and routes, which is less comfortable than many private transportation alternatives.

To better satisfy users, dial-a-ride systems have been initiated. Dial-a-ride systems can offer more flexible and comfortable door-to-door rides, chiefly by commercial vehicles and taxis [3]. To utilize the vehicles' passenger capacity, drivers can pick up other passengers before reaching the destination of the first customer. The authors implement a dynamic dial-a-ride system, which can re-optimize routes after picking up new customers during services. Therefore, this dynamic dial-a-ride system supports a many-to-many service—customers have different departures and destinations—and does not need booking in advance.

Web-based shared ride systems include *Google Ridefinder*[1], *Ride Now!*[2], *RidePro3*[3], *eRideShare*[4], or *Mitfahrzentrale*[5]. These applications provide textual Web interfaces to attract registrations of shared ride clients and hosts, and are maintained by local and

---

[1] http://labs.google.com/ridefinder.
[2] http://www.ridenow.org
[3] http://www.ridepro.net
[4] http://www.erideshare.com
[5] http://www.mitfahrzentrale.de

regional agencies with central databases. Mediated trips are usually regional or national travels, with inner urban travels generally not catered for. To request or offer a ride, users (clients and hosts) need to provide their home addresses, cell phone number, email addresses and requested trip details. Then the databases match requests and offers immediately, and feed back a contact list of potential shared ride hosts or clients. The choice is left to the users who can email or call their selections. Agencies need high-powered workstations, database servers and internet connectivity to run such an application. Personal computers or mobile devices with Internet connectivity are necessary as data terminals for the users.

2.2   Agent-Based Simulation and Shared Ride Applications

Simulation is an accepted approach to investigate the behavior of complex systems in general, and of traffic [4,5] and sensor networks in particular. Simulation allows to study information spreading in mobile ad-hoc sensor networks, MANETs [6,7], as well as in more specialized vehicle ad-hoc sensor networks, VANETs [8]. For the present problem, a geosensor network of heterogenous nodes with travel intentions, with autonomous travel behavior, and spatial restrictions to move, a multi-agent system is chosen for its simulation. Agent classes are designed to represent the different types of moving agents.

Several established agent-based simulation libraries exist that simplify modeling. Object-Based Environment for Urban Simulation, *OBEUS*[6], has been developed as a simplest implementation of geographic automata systems in .Net [9,10]. It is designed for urban processes and built in a cellular automata model with transition rules in form of functions. Entities in *OBEUS* can be one of two types, either mobile or immobile entities. In *OBEUS* no direct relationship is allowed between non-fixed objects. That means that *OBEUS* is not suitable for our simulation of locally communicating mobile agents. *Swarm*[7] is one of the popular libraries based on Objective C and has a Java wrapper. *RePast*[8] is a newer Swarm-like conceptual toolkit [11]. *RePast* is a free open source toolkit core in Java, while it has three implementations in Java, .Net and Python. Both approaches support to program multi-agent systems that are composed of larger numbers of agents with functions describing their behavior. *RePast* was used successfully for a large-scale peer-to-peer shared ride system simulation [12]. However, installing and using libraries is in itself a larger effort due to the constraints imposed by a given system design, so we decided to develop our system from scratch.

Previous research on a peer-to-peer shared ride system proposes a trip planning model on ad-hoc mobile geosensor networks [13]. The peer-to-peer system was designed to solve the problem of capacity limitations of centralized travel planning systems with large numbers of concurrent users in large dynamic networks and ad-hoc ride requests. The authors demonstrate that without a central service, shared ride trip planning with limited knowledge is possible and computationally efficient in a dynamic environment. They later implement this scenario with a simulation, in which clients with

---

[6] OBEUS can be downloaded from http://www.geosimulationbook.com

[7] http://www.swarm.org

[8] http://repast.sourceforge.net

transportation demand, and hosts with transportation supply communicate on a radio base to negotiate and plan trips in a continuously changing environment [1,14]. They design a mechanism for the negotiation process and investigates three communication strategies with different communication neighborhoods. Hosts are homogeneous, and clients are immobile in these experiments. The authors conclude that mid-range communication strategy in mobile geosensor network is both effective (leading to travel time comparable to complete current knowledge) and efficient (leading to less communication messages than those for complete transportation knowledge) compared to unconstrained or short-range communication.

3   Agents in Peer-to-peer Shared Ride Systems

Participants in peer-to-peer shared ride systems, to be modeled as geosensor network nodes later, are capable of perceiving their environment, of collecting information and making decisions, and of communicating where necessary. Particularly, peers are mobile, and some can move with other peers. In this section, immobile and mobile clients are identified, and three typical kinds of hosts (i.e., mass transit, taxis and private cars) with distinct economic and operational characteristics, in order to reflect better the properties of realistic shared ride systems in a simulation.

3.1   Clients

Real world clients have a desire to travel to their destinations and depend on rides from hosts. Immobile and mobile clients can be distinguished. Immobile clients rely completely on rides in order to move. Mobile clients can alternatively move on their own, but far slower than taking rides. The mobility of clients can depend on their preferences, their luggage, or their company (e.g., children).

Some clients might stick to preselected routes (e.g., the shortest) and only look for rides along their route. Alternatively, clients with a desire to optimize routes using cost functions such as travel time, number of transfers, or trip fares, will accept detours, as long as they promise to reach the destination for lower cost. For some clients, shorter travel times are more important than trip fares, while budget clients favor cheaper rides. Fewer transfers are more attractive to clients who appreciate comfortable trips, while scenic views would be a cost function (to maximize) for tourist clients. Frequently clients balance these factors with some subjective weighting. Furthermore, clients can have other preferences, such as for types of hosts, or for specific profiles of vehicle drivers.

Another factor to consider is the knowledge of the client. While the general assumption is that the client knows the street network for trip planning, it makes a difference whether the client knows also the mass transit network and time tables, or typical traffic patterns in the city (e.g., main streets experience more traffic than others).

3.2   Hosts in Mass Transit

Mass transit in a city includes buses, trams, trains, underground, and ferries. Generally, mass transit vehicles carry more passengers compared to other means of transport, although with less comfort and privacy. Travel fares are relatively cheap, especially with flat fare structures on longer distances, or with tickets that are interchangeably valid on various modes of mass transit. Often fares are charged by time only, regardless how long the trip.

Mass transit follows fixed timetables, typically with larger gaps between midnight and early morning and varying frequency over the day. They run on predefined routes back and forth, and passengers are only allowed to get on or off at stops. This means that mass transit does not provide door-to-door transport, and some areas are not served at all. Some means of mass transit run on their own line network, e.g., trains, trams and subway, or have reserved lanes, and are less affected by other traffic. This means that mass transit vehicles may be faster than street traffic bound vehicles.

3.3   Taxis

Taxis are more comfortable and convenient compared to mass transit. Taxis can reach every location in a city's street network, and can be called at any time of the day. Passengers can head directly to their destinations without compulsory intermediate stops or transfers. Detouring, change of destination, and stopovers are also possible during travel.

The main disadvantages of taxis are a limited passenger capacity, and correspondingly, a high trip fare. Normally, taxis have about four seats for passengers, but these are only shared by a group sharing the same trip. Taxis are charged by a combination of travel distance and time; sometimes a flag fall is added. This means that taxis are more suitable when time or convenience is more valued than money.

3.4   Private Cars

As hosts of shared rides, private cars are similar to taxis in some respects: they share the advantage of comfort, and the disadvantage of low passenger capacity. The difference is that private cars are owned by their drivers, and hence, are considered as private space, or proxemics [15].

Nevertheless, private car drivers may be willing to offer a ride if they get some incentives. But they are unlikely to serve clients off their route. Rather they pick up clients anywhere along their own trip, and give them a ride along their own route. Private car drivers may also have rigid interests and preferences in selecting clients, such as non-smoking clients, or clients of a specific gender.

Incentives for the car drivers could be nonmonetary, such as being allowed to use high-occupancy vehicle lanes with passengers on board. Even if they charge fees proportional to the traveled distance, their rates will be lower than taxi rates because the car drivers' interest is mostly sharing costs.

4   Communication in Peer-to-Peer Shared Ride Systems

Peer-to-peer communication in a shared ride application enables nearby agents to collaboratively solve the shared ride trip planning. To make optimal decisions, agents need to consider all transportation information. However, in dynamic traffic, agents have to make decisions with local knowledge only. This section discusses high-level communication protocols and strategies, the agents' negotiation mechanism and data collection in a peer-to-peer shared ride system, as they are proposed in the literature and studied in simulations [1].

4.1   Communication Protocol and Strategies

In a peer-to-peer shared ride system, the trip planning clients depend on transportation information from hosts. However, peer-to-peer communication for real-time decisions in dynamic street traffic enables only local communication strategies. This means an individual client may not reach or may not want to reach all hosts in the street network, and hence, has to plan a trip with local knowledge only. Nagel suggests that trip plans always include a start time, a start position, a destination and a sequence of nodes in between [5]. In shared ride planning, agents are additionally interested in the agents involved in the trip, arrival times, and travel fare. To enable negotiations between agents for trip plans, a communication protocol is designed for messages of the structure specified in Table 1. The details of the communication model and protocol are specified elsewhere [1].

Table 1. Message elements.

	Field	Type	Description
1	type	char	request r, offer o, booking b
2	route	[node]	requested or offered route
3	time	int	start time of the route in the message
4	agents	[int]	record of all identifiers of agents that transfer this message
5	speed	float	speed of the original sender of this message
6	fare	float	travel fare of the offered route

In a peer-to-peer system agents radio broadcast messages to their neighbors. Their radio range is limited according to the broadcasting technologies and the broadcasting power. Distant agents can be reached by forwarding messages (multi-hop broadcasting). For a peer-to-peer shared ride system the communication window—the synchronized time all agents listen and broadcast—requires to be long enough to accomplish a complete negotiation process, consisting of a request, offers, and a booking. So far trip planning with unconstrained, short-range and mid-range communication has been investigated in simulations. Unconstrained communication means that messages flood to the deepest agents in network, as long as agents are connected. Short-range communication means that agents only communicate to agents within their radio range (single-hop). In mid-range communication, agents forward messages for several hops. The negotiation process will be simulated for different communication ranges to investigate trip planning with different levels of transportation network knowledge. However, it is clear that the unconstrained communication strategy is not feasible in reality and used here only as a reference for the trip planning with (theoretically) maximum real-time information.

4.2   The Negotiation Mechanism

The mechanism to process the negotiations is shown in Figure 1. Clients initiate a negotiation by sending a request. Hosts respond with offers, clients make a selection, and the negotiation finishes with a booking made by the client. The three communication phases happen sequentially within one communication window. All requests, offers and booking messages are in the format of *message*, and are identified by *type* and the original sender in *agents*. After each negotiation, communication devices fall asleep to save energy, and agents move until the next negotiation process happens. Agents do not keep previous negotiations in memory. Therefore, there is no cancelation process integrated, instead booked rides are regarded as being canceled when no rebooking/confirmation happens in the following negotiation.

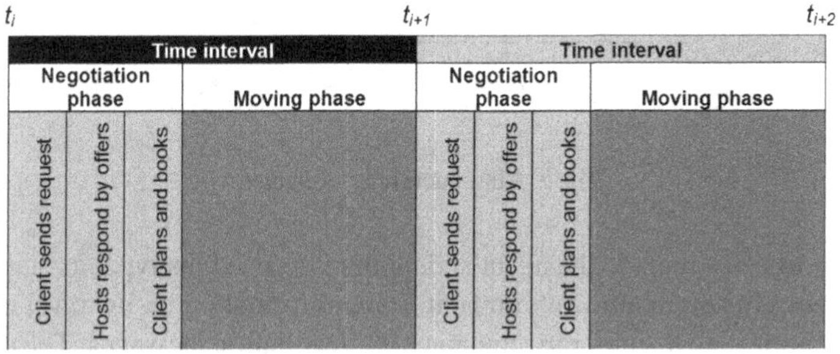

Fig. 1. The cycle of negotiations and movements within two time intervals

So far, only one client is generated in an individual simulation. All hosts serve this single client.

5   Formalization in a Multi-agent Simulation

This section presents a specification of a peer-to-peer shared ride simulation, with the types of agents (i.e., geosensor nodes) and their behavior as discussed above. The simulation is implemented in an object-oriented architecture using Java. Design details of the simulation model and related algorithms are elaborated by [16].

In our peer-to-peer shared ride system, agents have knowledge of their locations within the street network, negotiate with their neighbors for shared rides, make decisions according to their desires and intentions, and travel until the next negotiation takes place. Therefore, this system can be seen as a geographic automata system [10]: it has states, and state transitions, in particular the movements.

To implement geographic automata systems, Benenson and Torrens [10] suggest establishing a spatially restricted network with immobile and mobile agents, neighborhood relationships and behavior rules. Due to their interest on urban objects, such as buildings or residential addresses, they use a cellular network. In contrast, agents in shared ride systems move in street networks, and hence, we use a grid network to model a real street network, with nodes representing street intersections and edges the street segments. Agents run in the grid network, and negotiate in an ad-hoc manner for ride sharing.

5.1   Agent Parameters and Behavior

Agents are designed in a class hierarchy (Figure 2), because they all have some common features and behavior. These common features and behavior are identified and encapsulated in the base class *agent*.

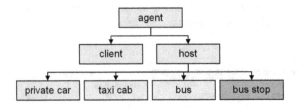

Fig. 2. Class hierarchy of agents

Common features include the agent's identifier, its speed, its type, its state, its location in the current simulation environment, some information on its travel plans, such as the destination, and a temporary container for negotiation messages. The travel route contains departure and destination, and for some agents the nodes in between. For investigation purposes, a second container stores details of booked shared rides. Common behavior includes how to move to the next node, how to listen to neighbors and how to obtain knowledge about current position and state.

The classes *client* and *host* are derived from *agent*, and have additional properties and characteristic behavior. Their states, travel routes and current position can change over time, but type and speed are constant within a simulation.

5.2   Client Agents

In the simulation, there are two types of clients: immobile clients, taking rides only, and mobile clients that are also able to move. The first type of client needs to be picked up from their location. The second type of client is able to move, which enables them to move to another location if they can get a ride there sooner. For clients, a (time-dependent) shortest path algorithm is needed for trip planning. The algorithm implemented is the heuristic lifelong planning A* algorithm [17]. This algorithm is adaptive to the dynamic traffic network. Given various cost functions (e.g., travel time or trip fare), this algorithm allows clients achieving different goals such as the quickest or the cheapest trip.

5.3   Host Agents

There are three kinds of hosts in this simulation: private cars, taxis and mass transit. These hosts vary in their mobility, in their routing flexibility, in their passenger capacity, and in their economic models. Implemented hosts have two modes to respond to requested trips: they can offer to share sections of their own travel plans that match with requests, or they can leave their predefined travel route and make a detour for clients. A third alternative—hosts offering their travel route ahead no matter how relevant this is to a request—would only increase the communication costs.

5.4 Quality of Trip Planning

Local communication provides limited knowledge for clients, accessing only the travel plans of nearby hosts for shared ride trip planning. This knowledge is limited from a spatial ('nearby', which depends here on the communication strategy: short-range, mid-range, or unconstrained) and temporal perspective ('now'). With this knowledge, clients in most cases can only choose sub-optimal trips. To investigate the consequences, an observer agent is designed in the simulation to enable a hindsight investigation of a global optimal trip. The observer is capable of monitoring the entire transportation network within the geosensor network. This global optimal trip can be compared with the client's trip to evaluate trip quality in the simulation.

6   Simulating Shared Rides with Diverse Agents

The specified peer-to-peer shared ride system simulation is tested for different types of agents. For the purpose of the test, travel time was chosen as the optimization criterion to look for the fastest trip. The simulation produces output in the form of text, which can be stored or visualized. Each result presented in this section summarizes 1000 simulation runs. For the experiments, hosts were parameterized according to Table 2.

Table 2. Parameter settings of various host types

	Type	Capa-city	Speed	Route	Detour	Fare rate	Others
1	*private car*	2	1	fix	FALSE	0.5	
2	*taxi*	1	1	variable	TRUE	1	flag fall is 1
3	*mass transit*	10	2	predefined	FALSE	-	schedule; one-off charge is 2

6.1 Global Optimal Trips Compared with Sub-optimal Trips

This experiment compares global optimal trips, computed posteriori for each simulation, with the client's trips made with two different communication strategies: mid-range (*comRange* = 3) and unconstrained (*comRange* = 20) in a grid network of 10×10 nodes (the radio range is generally set to one segment). In this experiment the client is immobile and follows the geodesic route from node $(3, 5)$ to node $(8, 5)$, that is, the trip is in the center of the network and has a length of five segments. Homogeneous hosts of type *private car* are generated at random locations and with random routes of twelve segments length. Host density, defined as the proportion of the number of hosts and the number of nodes of the grid network, is fix. Figure 3 shows the average travel times of trips made versus the average global optimal travel time for various host densities.

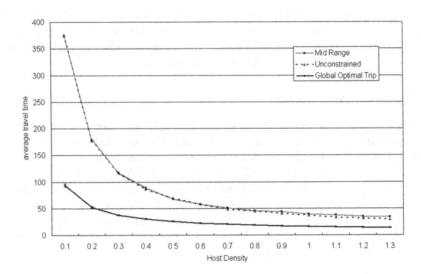

Fig. 3. Comparison of global optimal trips vs. sub-optimal trips realized by mid-range and unconstrained communication strategies

The experiment shows two significant results. First, a mid-range communication strategy is acceptable for all host densities; the unconstrained strategy, which is not feasible in practical applications, would improve travel times only marginally. Secondly, even complete current transport network knowledge as provided by the unconstrained strategy reaches only sub-optimal results, not considering future travel opportunities in time. – Since global knowledge is not accessible by clients, global optimal trips are not considered further in this paper.

6.2 Heterogeneous Clients Under Diverse Communication Strategies

This experiment compares the efficiency and effectiveness of diverse communication strategies: short-range (*comRange* =1), mid-range (*comRange* = 3) and unconstrained (*comRange* = 20) in a grid network of 10×10 nodes. There are four types of clients looking for the fastest trip: 1) an immobile client who sticks to the geodesic route, 2) an immobile client who is willing to make detours, 3) a mobile client who sticks to the geodesic route, and 4) a mobile client who is willing to make detours. Each client departs at $(3,5)$ and heads to the destination at $(8,5)$. Mobile clients have a walking speed of $v_c = 0.25$ edges per time unit, while the the homogenous host speed is $v_h = 1$ edge per time unit. The 72 hosts are all private cars.

Figure 4a shows the average time of shared rides by various clients, and Figure 4b shows the corresponding numbers of broadcasted messages. The experiment demonstrates again that (short-range and) mid-range communication delivers trips nearly as fast as unconstrained communication, for all densities of hosts. It also shows that short-range and mid-range communication produce much less messages than unconstrained communication. Furthermore, the client's ability to move and their flexibility to make

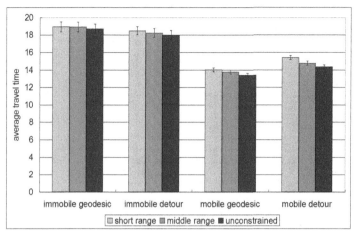

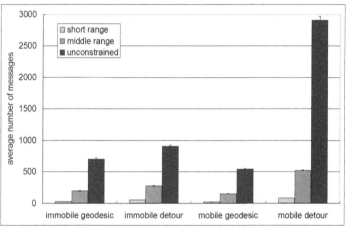

Fig. 4. Comparison of trip planning under three communication strategies

detours make a significant difference in travel time. Mobile and flexible clients, due to their increased choices, have advantages over immobile or inflexible clients.

6.3   Geo-routing Quality with Heterogeneous Hosts Using Local Knowledge

This section investigates a case with mobile, flexible clients in a transport network of various types of hosts in a grid world of 20×20 nodes. Mass transit is introduced as two bus lines (Figure 5), with one bus line partially overlapping with the direct route of the client. One new type of agent is the *bus stop* which is a static agent participating in negotiations and knowing the bus schedules.

Five experiments have been conducted, all with the same density of transportation hosts but with different proportions: 1) 144 private cars only; 2) 96 private cars and 48 buses (12 buses run in each direction of the two bus lines); 3) 96 private cars, 48 buses

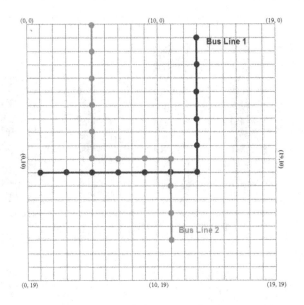

Fig. 5. The two bus lines in the grid street network

and 24 bus stops to help transferring bus travel information; 4) 96 private cars and 48 taxis; and 5) 48 private cars, 48 taxis, 48 buses and 24 bus stops. The average travel time and number of messages are shown in Figure 6.

The results show that the mix of host types has a significant influence on travel times as well as on communication efforts. In general, the presence of taxis in the network reduces average travel times, since once a taxi has picked up the client, the client travels along the shortest path. Buses also reduce the travel time because they are assumed to travel with double speed of cars (Table 2). Bus stops do not seem to have that importance, but this may be distorted by the relatively dense bus intervals in this experiment.

6.4   Mobility Models of Agents

Up to now, private cars and unoccupied taxis are traveling by random. In a more sophisticated agent mobility model host agents may have a preference of traveling *central* streets [18,19]. One of these models assigns connected segments of the grid street network to *named streets*. In this heterogeneous network of named streets, centrality was determined by *betweeness centrality* [20] and used to attract host traffic proportionally.

Clients aware of this behavior of hosts prefer to look for transfers at central street intersections, because there they have higher chances to find connecting hosts. To investigate this mobility model, experiments have to focus on the various behaviors of agents with different knowledge of the centrality in the street network. In the first experiment, the 120 hosts have no knowledge of centrality, and simply employ a random mobility model. Accordingly, the clients do not consider centrality either and follow strictly the graph geodesic between start and destination. In the second experiment,

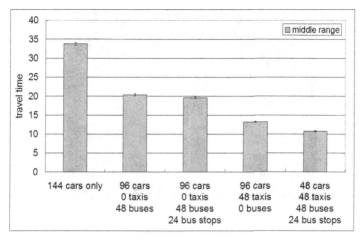

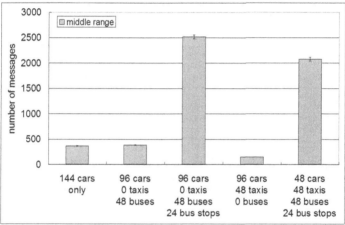

Fig. 6. Trip planning using local knowledge in multi-modal traffic

hosts have knowledge of centrality and adapt their mobility. Clients in this experiment still do ignore this knowledge and apply their traditional trip planning strategy. In the third experiment, finally, the clients consider centrality in their trip planning by favoring rides that end at central intersections.

Figure 7 visualizes the host distributions in the chosen named street network, where the hosts use the knowledge of central streets. The distribution of hosts is no longer equal, and the pattern shows the linear effects of long streets.

Then Figure 8 presents the results of the three experiments: the bars showing average travel times, and the points connected by a line showing the average number of messages. The smaller improvement of travel times between the first and the second experiment can be explained by the different qualities of the shape of the host routes: In average, the new mobility model leads to more elongated host routes than random movement, and hence, a single ride is in average more useful for the client. But more

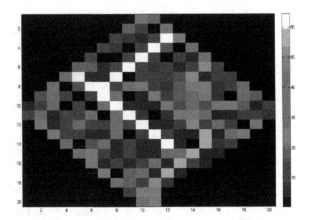

Fig. 7. Visualization of the host distributions, demonstrating a mobility model recognizing main streets and side streets in a grid street network

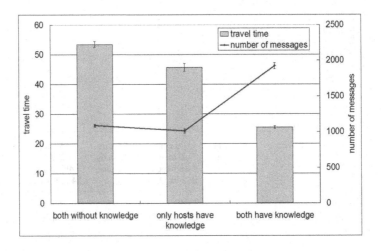

Fig. 8. Comparison of agents having different level of knowledge

impressive is the advantage for the client when adapting to the travel patterns of hosts, as shown in the third experiment. At the same time the numbers of messages increase because the clients are traveling through streets with more traffic.

6.5   Multi-criteria Optimization

The previous experiments were conducted on the assumption that clients want fastest trips. Nevertheless, in the real world people consider more factors when planning their trips. Other considered factors include the travel fare, the convenience (in terms of numbers of transfers), comfort, or security. More criteria make the planning of trips more complex: to make an optimal decision, people need to balance various criteria. This means the decision may not be optimal regarding a single criterion but good enough as a whole. In this section experiments are designed to investigate multi-criteria trip planning in peer-to-peer shared ride systems.

Three experiments are conducted, according to three types of client preferences: 1) clients prefer the fastest trip; 2) clients consider both travel time and fare; and 3) clients care about travel fare only. The third experiment has a trivial result: in the simulation, walking is always the cheapest way to travel, and the walking time is predictable, too. Therefore, to avoid the trivial case in this experiment, it is assumed that all clients are immobile, but would not mind making detours. Parameters in this experiment are set as before, with a host density in this case of 0.36, and mixed host types.

Multi-criteria optimization is implemented as a k shortest path algorithm [21] for the primary cost criterion, followed by a search for the optimal candidate according to the secondary cost criterion in this set of k candidates. It is assumed that clients choose travel time as primary, and travel fare as secondary criterion, and k is set to three in this experiment.

Figure 9 presents the three experiments, the bars showing average travel time, and the points connected by a line showing average travel fares. Multi-criteria trip planning (the second experiment) is neither fastest nor cheapest, but relative cheaper and quicker compared to the first and third experiment respectively. The average travel time of the multi-criteria optimization is only slightly above the fastest trip (note that the scales do not start at 0). The average travel fare, however, can be reduced significantly by taking this criterion under consideration as well.

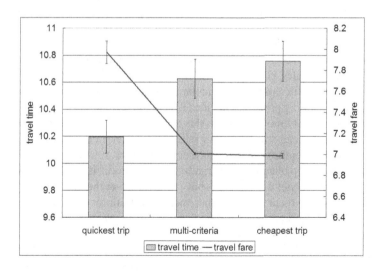

Fig. 9. Comparison of multi-criteria vs. single-criterion trip planning

7 Conclusions and Outlook

This chapter extends the context of geosensor networks—wireless mobile location-aware sensor networks—to an application in the field of transportation. A peer-to-peer shared ride system is presented composed of mobile geosensor nodes, which are either transportation hosts or clients. These nodes have heterogeneous properties, behaviors and interests. Since they are able to communicate over short distances with each other to look for or to provide rides in an ad-hoc manner, nearby nodes collaboratively try to optimize the satisfaction of the individual interests.

This chapter has been developed from previous research of a peer-to-peer shared ride system with an immobile client following a geodesic route, and homogeneous hosts that move on the basis of a random walking model. The previous research was extended by the introduction of mobile and flexible clients, various types of hosts, other agents, and more realistic mobility models. Finally, the clients were enabled to optimize their trips for multiple criteria.

Reviewing the results, multiple types of agents enrich the choices of clients, which leads to trips of lower costs. The largest impact has a system with mobile and flexible clients and all types of host agents, since it provides the largest choice. Mid-range communication still delivers trips of durations close to those from a (fictional) unconstrained communication range, but has much lower communication costs. Hence, the hypothesis has been proven. Since all experiments were parameterized by the density of hosts, and not by their number, one can expect that the observed results hold for longer trips as well, and also for other forms of street networks.

It is also shown that trips derived from local knowledge (of any communication range) may not be optimal from a global view. Better rides provided by distant hosts and hosts entering the traffic after the client has made a booking are always possible, and can be documented from a subsequent analysis of the simulation protocol. This problem can be approached by more intelligent wayfinding heuristics of the clients. Clients could, for example, learn from experience and use this knowledge in predicting chances of being picked up at specific nodes. For this purpose, a client could exploit a hierarchy in the street network, or known traffic counts at particular intersections, to assess potential transfer points in the trip planning process. This idea is being investigated elsewhere [19].

Although the mobility models used in this chapter are sufficient for the present simulation purposes, they can still be further refined to model more aspects of real traffic flow, such as cycles over the time of the day, or congestions. It is shown, however, that it brings advantages to trip planning if the random walker model is replaced by a more sophisticated mobility model where agents have knowledge of the main streets, and have a preference for using them. Hence, other meaningful improvements of the mobility models, and their consideration by a trip planning agent, are expected to show further advantages for the trip costs.

Multi-criteria optimization is essential for more intelligent wayfinding behavior. For example, clients may be interested to reduce their number of transfers and their trip travel time. The introduction of different fare structures, and the choice of the cheapest trip (or of a balanced cheap trip in a multi-criteria optimization) already tests economic concepts of a peer-to-peer shared ride system. The inclusion of more criteria requires another multi-criteria optimization strategy.

Another future extension of this system comes with admitting other clients into the simulation (*clientNum*>1). Then the passenger capacity of the hosts becomes a critical resource. Clients would compete with each other, which might recommend more booking ahead. But more aggressive booking strategies conflict with the hosts' interests of traveling with occupied vehicles, since travel plans are highly dynamic. Balancing these interests needs to be investigated.

References

1. Winter, S., Nittel, S.: Ad-Hoc Shared-Ride Trip Planning by Mobile Geosensor Networks. International Journal of Geographical Information Science 20(8), 899–916 (2006)
2. Miller, G.K., Green, M.A.: Commuter Van Programs - An Assessment. Traffic Quarterly 31, 33–57 (1977)
3. Colorni, A., Righini, G.: Modeling and optimizing dynamic dial-a-ride problems. International Transactions in Operational Research 8, 155–166 (2001)
4. Burmeister, B., Haddadi, A., Matylis, G.: Application of Multi-Agent Systems in Traffic and Transportation. IEE Proceedings in Software Engineering 144(1), 51–60 (1997)
5. Nagel, K.: Traffic Networks. In: Bornholdt, S., Schuster, H.G. (eds.) Handbook of Graphs and Networks. VCH Verlagsgesellschaft mbH, Weinheim, Germany (2002)
6. Nittel, S., Duckham, M., Kulik, L.: Information Dissemination in Mobile Ad-Hoc Geosensor Networks. In: Egenhofer, M.J., Freksa, C., Miller, H.J. (eds.) GIScience 2004. LNCS, vol. 3234, pp. 206–222. Springer, Heidelberg (2004)
7. Wolfson, O., Xu, B., Yin, H., Rishe, N.: Resource Discovery Using Spatio-temporal Information in Mobile Ad-Hoc Networks. In: Li, K.-J., Vangenot, C. (eds.) W2GIS 2005. LNCS, vol. 3833, pp. 129–142. Springer, Heidelberg (2005)
8. Sormani, D., Turconi, G., Costa, P., Frey, D., Migiavacca, M., Mottola, L.: Towards lightweight information dissemination in inter-vehicular networks. In: VANET 2006: Proceedings of the 3rd International Workshop on Vehicular Ad Hoc Networks, pp. 20–29. ACM Press, New York (2006)
9. Benenson, I., Aronovich, S., Noam, S.: OBEUS: Object-Based Environment for Urban Simulations. In: Pullar, D., ed.: 6th International Conference on GeoComputation, Brisbane, Australia, University of Queensland (2001) (CD–ROM)
10. Benenson, I., Torrens, P.M.: Geosimulation: Automata-based Modeling of Urban Phenomena. John Wiley & Sons, Chichester (2004)
11. North, M.J., Collier, N.T., Vos, J.R.: Experiences Creating Three Implementations of the Repast Agent Modeling Toolkit. ACM Transactions on Modeling and Computer Simulation 16(1), 1–25 (2006)
12. Tessmann, S.: Time Geography for Efficient Shared-Ride Trip Planning in Transportation Networks. Diploma thesis, University of Münster (2006)
13. Winter, S., Nittel, S.: Shared Ride Trip Planning with Geosensor Networks. In: Brox, C., Krüger, A., Simonis, I. (eds.) Geosensornetzwerke - von der Forschung zur praktischen Anwendung, Natur & Wissenschaft, Solingen, Germany. IfGIprints, vol. 23, pp. 135–146 (2005)
14. Winter, S., Nittel, S., Nural, A., Cao, T.: Shared Ride Trips in Large Transportation Networks. In: Miller, H.J. (ed.) Symposium on Societies and Cities in the Age of Instant Access, Salt Lake City, Utah (2005)
15. Hall, E.T.: The Hidden Dimension, Doubleday & Company, Garden City, NY (1966)
16. Wu, Y.H.: Agent behavior in peer-to-peer shared ride systems. Master thesis, Department of Geomatics, The University of Melbourne (2007)

17. Koenig, S., Likhachev, M., Furcy, D.: Lifelong Planning A*. Artificial Intelligence 155(1-2), 93–146 (2004)
18. Leigh, R.: Agent mobility model based on street centrality. Final year project report, Department of Geomatics, The University of Melbourne (2006)
19. Gaisbauer, C., Winter, S.: Shared Ride Trip Planning with Free Route Choice. In: Raubal, M., Miller, H.J., Frank, A.U., Goodchild, M. (eds.) Geographic Information Science, Münster, Germany. IfGI Prints, vol. 28, pp. 73–75. Institute for Geoinformatics, University of Münster (2006)
20. Borgatti, S.P.: Centrality and Network Flow. Social Networks 27(1), 55–71 (2005)
21. Yen, J.Y.: Finding the k shortest loopless paths in a network. Management Science 17(11), 712–716 (1971)

Author Index

Lecture Notes in Computer Science

Sublibrary 3: Information Systems and Application, incl. Internet/Web and HCI

For information about Vols. 1– 4730
please contact your bookseller or Springer

Vol. 4932: S. Hartmann, G. Kern-Isberner (Eds.), Foundations of Information and Knowledge Systems. XII, 397 pages. 2008.

Vol. 4928: A.H.M. ter Hofstede, B. Benatallah, H.-Y. Paik (Eds.), Business Process Management Workshops. XIII, 518 pages. 2008.

Vol. 4918: N. Boujemaa, M. Detyniecki, A. Nürnberger (Eds.), Adaptive Multimedial Retrieval: Retrieval, User, and Semantics. XI, 265 pages. 2008.

Vol. 4903: S. Satoh, F. Nack, M. Etoh (Eds.), Advances in Multimedia Modeling. XIX, 510 pages. 2008.

Vol. 4900: S. Spaccapietra (Ed.), Journal on Data Semantics X. XIII, 265 pages. 2008.

Vol. 4892: A. Popescu-Belis, S. Renals, H. Bourlard (Eds.), Machine Learning for Multimodal Interaction. XI, 308 pages. 2008.

Vol. 4882: T. Janowski, H. Mohanty (Eds.), Distributed Computing and Internet Technology. XIII, 346 pages. 2007.

Vol. 4881: H. Yin, P. Tino, E. Corchado, W. Byrne, X. Yao (Eds.), Intelligent Data Engineering and Automated Learning - IDEAL 2007. XX, 1174 pages. 2007.

Vol. 4877: C. Thanos, F. Borri, L. Candela (Eds.), Digital Libraries: Research and Development. XII, 350 pages. 2007.

Vol. 4872: D. Mery, L. Rueda (Eds.), Advances in Image and Video Technology. XXI, 961 pages. 2007.

Vol. 4871: M. Cavazza, S. Donikian (Eds.), Virtual Storytelling. XIII, 219 pages. 2007.

Vol. 4858: X. Deng, F.C. Graham (Eds.), Internet and Network Economics. XVI, 598 pages. 2007.

Vol. 4857: J.M. Ware, G.E. Taylor (Eds.), Web and Wireless Geographical Information Systems. XI, 293 pages. 2007.

Vol. 4853: F. Fonseca, M.A. Rodríguez, S. Levashkin (Eds.), GeoSpatial Semantics. X, 289 pages. 2007.

Vol. 4836: H. Ichikawa, W.-D. Cho, I. Satoh, H.Y. Youn (Eds.), Ubiquitous Computing Systems. XIII, 307 pages. 2007.

Vol. 4832: M. Weske, M.-S. Hacid, C. Godart (Eds.), Web Information Systems Engineering – WISE 2007 Workshops. XV, 518 pages. 2007.

Vol. 4831: B. Benatallah, F. Casati, D. Georgakopoulos, C. Bartolini, W. Sadiq, C. Godart (Eds.), Web Information Systems Engineering – WISE 2007. XVI, 675 pages. 2007.

Vol. 4825: K. Aberer, K.-S. Choi, N. Noy, D. Allemang, K.-I. Lee, L. Nixon, J. Golbeck, P. Mika, D. Maynard, R. Mizoguchi, G. Schreiber, P. Cudré-Mauroux (Eds.), The Semantic Web. XXVII, 973 pages. 2007.

Vol. 4823: H. Leung, F. Li, R. Lau, Q. Li (Eds.), Advances in Web Based Learning – ICWL 2007. XIV, 654 pages. 2008.

Vol. 4822: D.H.-L. Goh, T.H. Cao, I.T. Sølvberg, E. Rasmussen (Eds.), Asian Digital Libraries. XVII, 519 pages. 2007.

Vol. 4820: T.G. Wyeld, S. Kenderdine, M. Docherty (Eds.), Virtual Systems and Multimedia. XII, 215 pages. 2008.

Vol. 4816: B. Falcidieno, M. Spagnuolo, Y. Avrithis, I. Kompatsiaris, P. Buitelaar (Eds.), Semantic Multimedia. XII, 306 pages. 2007.

Vol. 4813: I. Oakley, S.A. Brewster (Eds.), Haptic and Audio Interaction Design. XIV, 145 pages. 2007.

Vol. 4810: H.H.-S. Ip, O.C. Au, H. Leung, M.-T. Sun, W.-Y. Ma, S.-M. Hu (Eds.), Advances in Multimedia Information Processing – PCM 2007. XXI, 834 pages. 2007.

Vol. 4809: M.K. Denko, C.-s. Shih, K.-C. Li, S.-L. Tsao, Q.-A. Zeng, S.H. Park, Y.-B. Ko, S.-H. Hung, J.-H. Park (Eds.), Emerging Directions in Embedded and Ubiquitous Computing. XXXV, 823 pages. 2007.

Vol. 4808: T.-W. Kuo, E. Sha, M. Guo, L.T. Yang, Z. Shao (Eds.), Embedded and Ubiquitous Computing. XXI, 769 pages. 2007.

Vol. 4806: R. Meersman, Z. Tari, P. Herrero (Eds.), On the Move to Meaningful Internet Systems 2007: OTM 2007 Workshops, Part II. XXXIV, 611 pages. 2007.

Vol. 4805: R. Meersman, Z. Tari, P. Herrero (Eds.), On the Move to Meaningful Internet Systems 2007: OTM 2007 Workshops, Part I. XXXIV, 757 pages. 2007.

Vol. 4804: R. Meersman, Z. Tari (Eds.), On the Move to Meaningful Internet Systems 2007: CoopIS, DOA, ODBASE, GADA, and IS, Part II. XXIX, 683 pages. 2007.

Vol. 4803: R. Meersman, Z. Tari (Eds.), On the Move to Meaningful Internet Systems 2007: CoopIS, DOA, ODBASE, GADA, and IS, Part I. XXIX, 1173 pages. 2007.

Vol. 4802: J.-L. Hainaut, E.A. Rundensteiner, M. Kirchberg, M. Bertolotto, M. Brochhausen, Y.-P.P. Chen, S.S.-S. Cherfi, M. Doerr, H. Han, S. Hartmann, J. Parsons, G. Poels, C. Rolland, J. Trujillo, E. Yu, E. Zimányie (Eds.), Advances in Conceptual Modeling – Foundations and Applications. XIX, 420 pages. 2007.

Vol. 4801: C. Parent, K.-D. Schewe, V.C. Storey, B. Thalheim (Eds.), Conceptual Modeling - ER 2007. XVI, 616 pages. 2007.

Vol. 4797: M. Arenas, M.I. Schwartzbach (Eds.), Database Programming Languages. VIII, 261 pages. 2007.

Vol. 4796: M. Lew, N. Sebe, T.S. Huang, E.M. Bakker (Eds.), Human–Computer Interaction. X, 157 pages. 2007.

Vol. 4794: B. Schiele, A.K. Dey, H. Gellersen, B. de Ruyter, M. Tscheligi, R. Wichert, E. Aarts, A. Buchmann (Eds.), Ambient Intelligence. XV, 375 pages. 2007.

Vol. 4777: S. Bhalla (Ed.), Databases in Networked Information Systems. X, 329 pages. 2007.

Vol. 4761: R. Obermaisser, Y. Nah, P. Puschner, F.J. Rammig (Eds.), Software Technologies for Embedded and Ubiquitous Systems. XIV, 563 pages. 2007.

Vol. 4747: S. Džeroski, J. Struyf (Eds.), Knowledge Discovery in Inductive Databases. X, 301 pages. 2007.

Vol. 4744: Y. de Kort, W. IJsselsteijn, C. Midden, B. Eggen, B.J. Fogg (Eds.), Persuasive Technology. XIV, 316 pages. 2007.

Vol. 4740: L. Ma, M. Rauterberg, R. Nakatsu (Eds.), Entertainment Computing – ICEC 2007. XXX, 480 pages. 2007.